BOOK OF SECRETS

By Fennel Hudson:

A Meaningful Life
A Waterside Year
A Writer's Year
Wild Carp
Fly Fishing
Traditional Angling
The Quiet Fields
Fine Things
A Gardener's Year
The Lighter Side
Friendship
Nature Escape
Book of Secrets
The Pursuit of Life

Fennel's Journal

No. 13

BOOK OF SECRETS

By

Fennel Hudson

2018

FENNEL'S PRIORY LIMITED

Published by Fennel's Priory Limited

www.fennelspriory.com

Limited edition magazine published in 2017
Abridged eBook published in 2017
This extended edition published in 2018

A CIP catalogue record for this book
is available from the British Library.

ISBN 978-1-909947-58-0

Available to purchase in other formats at
www.fennelspriory.com

Designed and typeset in 12pt Adobe Garamond Premier Pro.
Produced in England by Fennel's Priory Limited.

CONTENTS

STOP – UNPLUG – ESCAPE – ENJOY

This book, and the series to which it belongs, is about freedom. It's also about the adventures to be had when pursuing one's dreams, developing and communicating one's self, and striving for a slow-paced rural life. It's your opportunity to take time out from the stresses of modern living, to stop the wheels for a while, unplug from the daily grind, escape to a quiet and peaceful place, and enjoy the simple life. Because of this, I'd like you to read it in a distraction-free and relaxing environment: your 'safe place' where you can savour quality time and, if possible, delight in the beauty of the countryside.

That's why the book is pocket-sized, has a waxy cover and is printed using a special waterproof ink. It's designed to be taken with you on your travels. Don't store it in pristine condition upon a bookshelf; allow it to reflect the adventures you've had. Use a leaf as a bookmark and annotate the pages in the spaces provided with ideas of how you will honour your right to 'never do anything that offends your soul'.

The more mud-splattered, grass-stained, and pencil-scribbled this book becomes, the more you've demonstrated your ability to pursue a contented country life. So go on: live your life, be authentic, and always remember to 'Stop – Unplug – Escape – Enjoy'.

"Your visions will become clear
only when you can look into your own heart.
Who looks outside, dreams; who looks inside, awakes."

Carl Gustav Jung

INTRODUCTION

Welcome to the book that links each title in the Fennel's Journal series together. With 14 titles in the series (apparently) and 14 core chapters within this book, *Book of Secrets* is the 'one book to bind them all'. Each chapter is a prequel to the Journal to which it refers, thus providing continuity from one Journal to the next. *Book of Secrets* can therefore be read as a book on its own or, ideally, in stages where you read a chapter and then go back and read the associated book. I trust this will add meaning to the Fennel's Journal story and enhance your appreciation of the adventure it records. It's been about a life rebuilt, beginning when I had nothing and ending when I had more than the sum of my dreams. From 2006's *A Meaningful Life* to 2016's *Nature Escape*, it's been a decade of dreaming, writing, striving and living.

This book, my 'Book of Secrets', was previously reserved for my eyes only. It's shared with you now to aid your understanding of why I themed each book as I did and why I'm so passionate about each subject. The passion brings balance and life.

Life *should* be purposeful and balanced. But for balance to exist, two opposing forces must exist in harmony. Not easy when one side seeks to dominate the other. Fennel's Journal celebrates and gives weight to one of these forces: the good life. It's a quiet life, a slow-paced life, a natural life. Its roots, stems, leaves and blossom are outdoors in the countryside. Writing the series helped me to focus on the beautiful, cherished, humorous and serene. Hence why Fennel's Journal is deliberately sentimental, quirky, eccentric and indulgent. It's weatherworn and organic, natural, passionate, romanticised and unashamedly escapist. Each book seeks to transport us to a tranquil place, somewhere in the slow lane of contented country living.

Fennel's Journal was written to help keep the balance. As you may have gathered, for there to be so many books in the series, there was an equally large opposing force. Another side to the story. One kept secret, until now.

Writing Fennel's Journal enabled me to survive and overcome extreme adversity. For you to understand the extent to which it balanced my life, I will have to share a paragraph of darkness with you. It's taken directly from my 'black book' – the one that contains my saddest thoughts – but which, in this passage, provides a ray of hope:

"Fennel's Journal has been my lifeline. It's kept me focused and strong during a time when I have: lost my

sense of self; been betrayed and belittled by those closest to me; been abandoned at the altar; suffered crippling debt and fought insolvency; questioned my faith (when I nearly lost my wife and unborn child due to complications in pregnancy); sacrificed my happiness and health to clear my debts (by working hundred-hour weeks for nine years); endured physical and psychological abuse in the workplace (which gave my 39-year-old colleague a fatal heart attack); had two years of police protection (while violent neighbours vandalised my home and terrified me and my family); battled intense anxiety that kept me indoors for longer than I'll ever admit; been torn apart by high and low moods; suffered three breakdowns; narrowly escaped being sectioned; been medicated to the point of losing all internal thought; and cried as I held the hand of my closest friend while his life and body was destroyed by illness. Ten years in all, where it was barely possible to know what was real and what was not. 'Life' was almost too intense to comprehend. But I never gave up. I knew that if I kept on writing my Journal, in a way that entertained and inspired my reader, then he or she would be there with me – helping me along, reminding me of the good times."

Now you know my big secret: that while Fennel's Journal tracks my pursuit of an uncompromisingly tranquil rural life, the other side to the story has been intensely challenging and, at times, heart-stabbingly miserable. Hence why writing Fennel's Journal has enabled me to stay healthy. It's maintained my positivity and encouraged me outdoors to do the things I love. It's kept me true to myself, in touch with my friends, and always appreciative of what's important in life. It's half-a-million words of hope; narrative therapy that cements my absolute belief in the purity, honesty, and open-armed loveliness of life. It casts its eyes away from the savagery, sabotage and scarring brought about by others, and the traumatic twists of fate that sometimes befall us. It chooses its perspective and interpretation of reality. Just like we do when experiencing life.

The learning? That when you consider the 'darker side' backdrop to Fennel's Journal, you should never, ever, feel sorry for me. Anything I endured was either of my making, inertia or interpretation. I could have acted or reacted differently or viewed things in a better light. Instead I chose my stance, which influenced my perspective of life. This book, therefore, is not a Book of Sympathy. It's a Book of Secrets. If anything, I want it to provide a fuller picture so that you're encouraged by my journey and energised by my dogged determination to pursue my dreams. If I can do it, then so can you. A life *can* be rebuilt.

Writing Fennel's Journal, with its focus on the positive, the wonderful and the uplifting, has been my salvation. I've also found solace and meaningfulness in natural places that quietly weave their magic and protect us from life's hardships. One such place is a boathouse in Oxfordshire. It's surrounded by beech trees that sing in the wind, sounding like waves crashing on a shingle beach, and is beside one of my favourite trout lakes. I can sit outside the boathouse, on a lovely hand-carved picnic bench, and watch trout rising as I contemplate my good fortune. It's where I've written several of my books. The boathouse is my sanctuary, my writing retreat. It's where I can escape my worries to be, in the moment, carefree and content. Hence why it features on the cover of *Book of Secrets*. It's my special place, my secret place, my place of being.

Life now is sublimely good. Fennel's Priory is my everyday focus and reward, so the time is right for me to share my once-private writing with you. I know that, if you've read all of my books to date, *Book of Secrets* is as much yours as it is mine. It's presented in its near-original state. The only changes I've made are swapping the words 'my reader' with 'you' to make it more personal.

These were my secrets. Now they are ours to keep. Thank you.

Stop – Unplug – Escape – Enjoy

What's the other side of your story?
Is it known, or secret?

I

THE WEDDING SPEECH THAT NEVER WAS

The Fennel's Journal series of books traces a ten-year journey where I rebuilt my life, making good for a five-year period where I'd abandoned my true self and lived irresponsibly. It was a period that began severely, with consequences that damaged my health but improved my worldview. The story, therefore, has a dark beginning. But for everything to make sense, you need to know what came first, before A Meaningful Life *was written.*

There was another time, another life, when things were different. When I thought differently. When I believed that my life would never change. But it wasn't to be. That life ended. Suddenly and with great impact on everything I valued, it ended. I faded into darkness, knowing that silence was my new best friend.

'Death', I'm pleased to say, came early for me. I was 29 years old and grateful to escape the pain of suffering. Suffering, I'm keen to mention, that was of my making. I sank into a black void where once my heart had beat. And then I sat there, grimacing like a traffic warden on overtime, enjoying being miserable.

Of course, I wasn't *really* dead. It just felt like that. I'm a creative person. A writer. An artist. And yes, you've guessed it, a complete and utter 'drama queen'. I was broken and hurt, for sure, but not *dead.* I just existed without hope, not thinking or doing, while drifting in the nothingness of depression. And then I started to forget. With each lost memory I gained a new degree of happiness. So, I buried my memories and bitterness deep within. Ultimately, I forgot why I was so upset. All I would allow myself to remember was that I'd died much earlier. *Five years* earlier, to be precise, when I'd chosen to abandon the things that made me 'me'.

It was spring 1998. I was enjoying the pinnacle of relationships with my fishing and writing mentors; I'd just been asked to write my first regular column in a magazine and was finding a foothold in the world as an estate gardener. I'd put my trust in people and had adopted the nickname they'd given me. 'Fennel' was my way of knowing and communicating who I was, why I existed, and with whom I should travel. And then I made a mistake, at least in their eyes. I started to get good at being 'me'.

The impact of my success played out in two ways: the first involved some official advice given to me by a bank manager; the second took the form of confidence-destroying ridicule from my so-called friends.

The bank manager, concerned that I was regularly

exceeding my overdraft facility, told me I was too poor and, in so many words, too *pathetic* to ever amount to much. He withdrew my overdraft facility, put a hold on further borrowing, and advised me that – given my current trajectory – I'd never be able to afford a house, a car, or new clothes. I was, as he put it, "building a life of dreams on the vapour of nothing". I was instructed to think hard about my lifestyle and make some radical alterations to improve my cash flow. Tough love, but not as damning as the pressure and humiliation that came from my friends.

I'd gone on holiday to France in April, fishing for carp while I was there. There was nothing wrong in this. But in the UK, up until three years earlier, it had been illegal to do so at this time of year. (A 'closed season' had been in place to protect the fish while spawning.) According to three of my closest and most respected friends, the rules that had been in place in the UK should now apply in France. Consequently, I was to be punished for breaking their rules. How did they do this? I was forced to stand up in front of my wider group of friends while my three chums mocked and belittled me until I could no longer contain my tears. Harsh treatment given that it was perfectly legal (but not traditional) to fish for carp in April, and that I was a sensitive and impressionable twenty-three-year old. But my actions were an opportunity for them to knock me back and remind me that I was standing a little too tall in my boots. Always one to shy away from confrontation, I retreated into the sidelines; nursing my wounds and wondering whether the people I'd trusted were my friends at all.

How could someone in authority believe that I was worthless and likely to amount to nothing? Why did my mentors take so much pleasure in belittling me? Was I so terribly off-track and misguided? But – hundred-dollar-question – was I so feeble as to believe them and accept my fate? It forced me to question my lifestyle, values, friendships and character. As I did so, I grew angrier and more hateful. Fuelled by white-hot rage (that motivates me to this day), I set about proving these people wrong. I quit my much-loved gardening job and gained employment at a garden centre – stocking shelves for a living; I enrolled at night school, seeking to learn the ways of business; and forgot my values as I went in search of fortunes laid beneath a hangman's noose. I quit my Fennel identity, reduced my writing output, distanced myself from my friends, and rapidly spiralled out of control.

Being hurt by people we respect and trust can trigger rejection, depression and *furious* rage. It caused me to become a success-driven workaholic, obsessed with perfectionism and near-impossible levels of output. Regardless of what was expected of me in my job, I worked a twenty-hour day, seven days a week. The harder I worked, the more I was able to ignore recent events. Staying busy buried the pain of betrayal and gave me new levels of seemingly superhuman energy. I didn't need to sleep, I barely ate; I cared for nobody but myself. I felt like I was exploding, sending shockwaves through the world, but I was in a whirlpool of inevitable implosion.

Fast forward five years and I was barely recognisable in appearance, actions or words. I'd put on three stone due to alcohol abuse; was intensely focused, but with yellowing eyes that had huge black pupils; was writing more than ever – in my new job as a marketing copywriter – but never under my name; and was fishing with the most untraditional tackle and tactics. I was deluded, arrogant and foolhardy. I'd convinced myself that I was the best in the world at 'doing what I do'. Rather ironically, amidst the chest beating, I was actually hailed as being the best in the UK.

It doesn't take much to feel like a big fish if you choose a very small pond in which to swim. Mine was more of a puddle: horticultural marketing. More specifically, I wrote the text for plant labels. Yet, if you'd heard me boasting, you'd think that I was writing speeches for the Prime Minister.

My bravado, in retrospect, was sickening. I was an opinionated, self-centred and wholly unattractive prima donna intent on bounding over anyone or anything in my path. My moods were becoming unpredictable: defined by bouts of unexplainable frenzy and creative brilliance, followed – once I'd completed a task – by mind-shattering and gut-wrenching depression. Yet, amongst this bipolar chaos, I'd found a girlfriend who liked me. We'd been together for two years and were soon to be married. We weren't right for each other, encouraging the worst from each other's characters, but we lived for the moment and enjoyed wild times. We were a reckless pair who didn't know better. Until it was too late.

Saturday the 13th September 2003 should have been the biggest and happiest day of my life. I was due to marry my soulmate. Or so I thought. But it wasn't to be. Two weeks before the big day, I discovered that my fiancée was seeing someone else. They were planning an alternative future together, where they would take everything of value and I'd be left with nothing but a broken heart and debts equivalent to three times my salary. I confronted my fiancée; still hoping for a future together, but it wasn't to be. The wedding was cancelled and, two days later, I returned home from work to find the house emptied and my fiancée gone. I was left with twelve maxed-out credit cards, a car loan, furniture loan, overdraft, and a mortgage worth 125% of the house value. Oh, and an envelope containing the wedding speech I'd written. It was unopened. It has remained so – until today.

I have now read the speech and been taken aback by what it says. It shamefully contradicts everything I hold dear and is an anti-manifesto for the life I seek. Because of this, I have committed to living a meaningful life. In doing so, I feel like I have returned to being the real me – the one who existed before the events of 1998 – and put right the 'near miss' of what could have been if my wedding speech had been read.

I am reborn: the gentle one, who values time outdoors more than money, who's known for his virtues as a loyal friend, who fishes traditionally on his terms, who senses the delicate changes of the seasons and the fluttering, pattering, roaring magnificence of nature. I am a writer, the one who places his heart – not bravado – upon the page. I'm the one with friends who understand, even if they are a tenderly held few. I am the man I have chosen to be. Not because someone made me, but because I exist deep within. I am *revealed.* True to myself. I am, once again, Fennel.

I am not the person who wrote the wedding speech. If you don't believe me, read it for yourself:

"Dear friends and family. I know that there's an established way of giving speeches like this, with formalities acknowledging my wife and thanking my in-laws for paying for everything, but I'm not going to do any of those. Because it doesn't matter. It doesn't matter to me that this is my wedding day. It's just a day between yesterday and tomorrow in which we'll have some fun. It's also a day when I'm not working or earning. Is it a good thing? I don't know. It's not costing me anything, so does it matter?

No. It does not matter. It doesn't matter whether I stand here and speak the truth. You'll decide what you wish to hear. But will I have patience to wait for you to hear my words? I doubt it. It doesn't matter. Some of you are friends with me because you choose to be; some of you don't have a choice. You might be my colleagues. You might be my family. I must have influenced you somewhere? Have faith, they say. 'There's meaning to all this.' I used to be weak. But now I'm strong. And I'm stronger with my wife. Together we are strong. We will survive, no matter what. Because nothing else matters. We 'exist'. That's the point. Our job is partly to remain, but mostly to excel. 'Always shine brightly. Be brighter every day,' they say, 'Darkness is only dark in the absence of light.' Ha. Who could be so weak as to let the darkness in? We are the light.

I used to think that there was more to life than excellence alone. Lesser things used to be important. When I was a child I would love nothing more than time outdoors amongst wildlife. How sad I was, thinking that Nature could be seen as a person in her own right, or that there was some guiding hand controlling the things that can only be wild. I used to like gardening, but it paid terribly. How can we enjoy something that doesn't make us rich? I used to go fishing, but what a criminal waste of time it is. Days and nights when one could be working hard to change the world. I used to think that it was important to be sensitive, 'in touch with one's emotions', feeling something in items that gave them meaning. But there's no such thing as beauty. Only attraction and lust. And that can be animalistic. I used to think that things could be free, at least things containing value other than their worth, but everything and everyone has their price. I used to think that there was value in the past, in traditions and customs. But the past is dead. Who wants to make love to the dead? That's sick. Stay alive and screw the living. Live only for today. Tomorrow doesn't exist, either. I once dreamed of and for the future, of cosiness at home – maybe a timber-framed cottage with an open fire, leaded windows, roses in the garden and children running about the lawn – but home is where the heart is. My heart is with my wife, wherever she may be. Mostly I find her drunk in nightclubs, so that's my home. It's where I go when I've finished work. It's my out-of-hours calling – two hours of partying before I crash for my two hours' sleep. There's a secret to not sleeping: it's called a packet of caffeine tablets crumbled into a half-pint of vodka. It's a great way to sleep while still awake. I've ditched the vodka for today, but not the pills. I'm properly awake, fuelled by twenty heart-pounders that remind me that this is my wedding day, the day that marks another success in my life. We exist for the cause – not to do good for others, but for ourselves. We cannot trust others. Friendship is folly. We cannot let their words guide us or help us. We have to do this on our own, whether we're married or not. I stand before you as a symbol of success. I am the best, and I've got what I want.

My name is Nigel. I got married today, for today.

There's no other reason, as nothing else matters. This is my speech. So hear my words: make only a promise to yourself – it's the only way life can have meaning. There's no waterside beauty, no point in writing about it, no quiet fields that seek to heal us, no fine things or sentiments, no 'smelling the roses', no lighter side that promises happiness, no true friends, and no escape to a better place. There are only actions and exploitations. Live fast. Live without consequence. There is no tomorrow."

There you have it: the wedding speech that never was. Never have I read anything that so opposes my values and beliefs. It is testament to the extent to which I had fallen. No wonder I was jilted and made to pay, and why the sense of disbelief hurt so much.

I've been away for too long. To move forward and build a new life, I will have to make a promise to others so absolute as to ground my future thoughts and actions. And, most importantly, I must always seek to lead a meaningful life – on my terms, doing good for others. I will correct the shortcomings of my past and make good for the future. One day I will make a very different sort of speech, highlighting my pursuit of life. To get there, I'll need to stay true to myself and write regularly. One word at a time, capturing all that is real.

My name is Fennel. This is my Journal.

II

SEEING THE LAKE FOR THE TREES

Sometimes we need to be alone. Away from others, from distractions and from our 'other selves' that exist amongst certain circles. We need time to ourselves to hold and nurture our child within. Time away, from places and things. To unravel and come undone; to deliberately lose the plot, so that we can piece together a new story for our lives.

In the quiet of our minds, we heal.

I'm happiest in my own company. Not that I don't value friendships; quite the opposite. But I relish opportunities for quiet times alone. It's my default position, allowing me to reboot and re-energise. Without it, I pale into a faint version of myself. Better for me to retreat every now and then, to think, feel and drift, to come back stronger. A quiet walk might be enough, especially at dawn or dusk, or travelling to remote places such as the depths of a wood or the peak of a mountain. There, silently exploring, I am loudest. It's when I can most hear the truths within.

A lion is likely to roar when released into the wild. So it is with our callings. 'Time away' enables us to remember our 'why'. Passions are forged with wild resolve. They burn magnesium bright, shining the brilliance of truth onto a hungry soul – that we are part of nature and it is part of us. Indeed, it is all of us. Organic beings, we are born into this world and should remain connected to it. Yes, time spent alone sure helps us to see and think clearly – even if it does lead to some hippyish beliefs.

I've spent a large part of my life 'searching'. Ever since adolescence, I've been trying to understand and shape my identity so that I know where I belong in the world. It's not easy for someone as sensitive as me. I forge a 'pure' direction and then get knocked off track by some unexpected incident. I was doing okay until 1998, then things went awry. Now, nine years later, I'm wondering whether I've spent too long looking backwards and not enough time focused on the road ahead. But to know where we are, we have to know where we started and where we've been. Hence why I did my best to retrace my footsteps to my true self last year. But there's a secret. As I write these words (it's January 2007), *A Meaningful Life* only exists as notes and letters shared with my friends. There are eight thousand words at best, only a quarter of a book. It's my intention to expand and reshape the material and its message. But I don't know how to do this, because I haven't yet found my voice. I need more time to think about the words and practise my speech. I need time alone, to remember who I am and why I exist. I need to remember my calling. Above all else, I need to figure out how to honour The Promise.

I'm going to do something radical. I will spend as much time as possible this year being alone outdoors. Being close to nature is the calling that sounds loudest. It's the one thing I know to be true: that I love 'the wild' more than in any other place. I am a child of nature, in need of green therapy.

I'm also going to change my job. I've been working for the past three years as a marketing copywriter and team leader for an environmental company. It's been interesting work, but the pressures of managing the inputs of others rather than doing all the writing myself have proved too stressful. I'm not ready for this responsibility; I'm not yet strong enough to go back to 'that' world of perpetual stress. I've had my empire and seen it crumble because I too was crumbling. I don't want that again, for my team or me, so I'd rather do something quieter. With luck, I'll get a job as a 'pure' writer: one who's paid by the word rather than tied to a salary and all the politics that go with it. If I do, then where I work is unlikely to matter to my new employer. If I play things carefully, I'll be able to avoid the office and submit my writing via email. 'Working from home' would enable me to spend lots of time outdoors. I could work from a laptop computer powered by a stack of spare batteries, typing my target word-count per day, then spending the rest of the day writing my journal by hand and soaking up the wonder of nature. It would give me the time and lifestyle I seek to figure out what's been going on in my life and how to forge a new one for me and my new Mrs-H-to-be.

Mrs-H-to-be, whom I started dating in 2004, is infinitely better than the one I lost. She's supportive of the real me and my plan for a better future. In fact, she's the one who gave me the idea of spending a year in the wild (knowing that, to grow roots deep enough to feed my dream, I would need to remain 'grounded' for a long time). The complication, however, is that she and I are getting married in September. So while I will be away writing and 'living the life' of a rural tramp, she'll be planning a wedding without me. Of course, while I'll inevitably tell everyone that I'm living wild *all* the time, I'll actually be coming home or going into the office for a day or two each week. I'm not going to be *totally* unwashed, unpaid, or unloved. I'm going to be keeping things in balance. But my focus will be on my time in the wild, doing my best to become the best and only version of myself. So that when I say, "I will" in church, I'll know what and whom I'm willing to my other half. I'll also be committing to being this person before God. I am, after all, spiritual man who seeks to remain true in the eyes of those who matter most. But to get there, I need time alone to sort my 'self' out.

You could suggest that this 'waterside' year is a

new start and proper beginning for Fennel's Journal. While *A Meaningful Life* is the first in the series, 2007 will be when I start reshaping my future – albeit in an isolated fashion and seemingly shirking from my obligations to others. But hey, I'm an introvert. It's in my nature to be selfish, self-centred and antisocial. Blame nature, not me, for my wanting to spend as much time alone as possible.

My plan, therefore, is to enjoy proper 'time out': an analogue existence with no phone or outside distractions, where I will live in near-isolation. Sure, I'll be popping home and into work, and I'll take regular walks to the nearby village for supplies and to post my letters; but mostly it will just be me – camping somewhere remote, ideally in the life-giving presence of water. (As an angler, water replenishes my soul more than anything else.) I'll fish, forage, hunt, write, read, appreciate my surroundings, be grateful, look after the campfire, and do my best not to be freaked out by anything unexpected that I discover. It will be a Waldonesque existence, first championed by Henry Thoreau, where I'll live free from the hands of a watch and be, for nearly a year, totally free.

Freedom, defined by autonomy, is what I most crave. I want to do what I want, when I want, choosing to be whomever I want. Because I know that when this year is over, the pressures of life will inevitably wrap their claws around me once more – doing their best to nudge me towards a less meaningful existence. So I'm going to enjoy the simple life while I can, using it as the reference point for whenever I need to reboot myself in the future. It will be my first nature escape, but unlikely to be my last. Here's praying that it will be the lifeline I seek, now and forever. Amen.

Prequel to A Writer's Year

III

FINDING ONE'S VOICE

I've just returned from a nine-month stint living alone beside a lake. Spending so much time in the wild reaffirmed my love of isolation, but it made me yearn to be home. I realised that while I've been indulging my escapist dreams, I've been hiding from my problems. I had to come home and, with Mrs H holding my hand for support, confront the world.

I wrote on the back of my last notebook at the lake: "It's better to run towards something good, than away from something bad". I knew, deep down, that I'd been running since my bank manager and so-called friends crushed my dreams of a simple life in 1998. It was time for me to stop running. The best way for me to do this, I concluded, was to run towards the thing I most desired but most feared: being a successful full-time writer.

I've worked as a corporate copywriter for years. Writing on behalf of others now seems like a cowardly way of writing. I need to speak out *as me* and be known for being something more than a corporate wordsmith. Therefore I've decided to write professionally under my name (at least, the herb-based pen name of my choosing), ultimately to make my living from it. Writing copy on behalf of organisations or individuals is no longer good enough. I need my voice to be heard, sharing my message, on my terms. That's the journey I'm going to take. It's my new path, my new mission, my new calling. I will be known for encouraging others to 'Stop – Unplug – Escape – Enjoy', to explore the countryside and their inner thoughts so that they may see and appreciate the world in new ways. Everything henceforth will support this goal. I *will* become a successful writer. And about bloomin' time, too.

My decision is timely. I've received a letter today from my old friend Mike 'Prof' Winter. It says: "Publish a book before you're too old to read it without glasses". I know what he means and sense the frustration in his words. I am, after all, the person he once described as being the most naturally gifted writer he'd ever met. (In 1998 he'd seen me write, edit, proofread and finalise 6,000 words of my *Wild Carp* book in just four hours.) He'd received my handwritten 'journal' from its beginnings in 1996 and always replied with encouragement for me to keep writing.

As a schoolteacher and journalist, Prof has a gift of knowing how to nurture a talent and share a story. He's rightly frustrated by my insistence on keeping my work to just a small circle of friends. But I've been unready to have my work scrutinised by those I don't know. Too shy and modest, for sure, but mostly too scared. Prof is the person who, more than anyone else, has encouraged me to keep writing regardless of what others – or even myself – might think of the output. "You're a perfectionist, Fennel," he'd say. "Your writing will *never* be good enough for you; but it's plenty good enough for the rest of us. So put it in front of editors and publishers. Get your work in print!"

This year will be when my focus and confidence as a writer will change for the better. I'm going to champion the writer's craft and start building a body of work for publication. One day I *will* make my living from creative writing. 'Success' to me, however, is not just commercial: it's about reaching out and connecting with a reader in a way that has lasting and meaningful impact. One's writing, after all, is a legacy that provides a snapshot of our world for those who inherit it from us. But if the writing's strong enough, it might also enhance the lives of others today.

What will be my legacy? That I'm a rural lifestyle author, the one who writes openly about the beauty of the countryside and its abilities to help us escape from the pressures and challenges of life. I'm on a journey, building a brighter future that's inspired by the past, attempting to encourage others to get outdoors and follow their dreams. (If a shy, burned out, penniless, traumatised depressive such as me can do it, then so can anyone who has a glimmer of hope in their heart and fresh air in their lungs.) But that's the big picture. What about the role that writing plays within the story?

Writing is a way to speak without speaking; to connect without meeting; to relate upon the page and bond between the covers. It's an act of love that enables the writer to reach out into the world, permanently.

Words have the power to transform people. I was taught at marketing school that writing is a form of communication; that it seeks to do one or more of six things: *inspire*, *inform*, *educate*, *entertain*, *influence*, or *persuade* the reader. Its purpose, so my textbooks said, is to get the reader to take action. For me, my hope is that you will see the world differently after reading my books. Hopefully you'll then engage with your surroundings in new or better ways. Maybe, if I don't change or enhance your perspective or understanding, I'll reinforce your beliefs. Or I might simply annoy you. If you don't like my work, then that's fine. It's not for you, so you can move on. What's for me to fear? Merely the risk of not sharing my work or having nothing to say. As a BBC journalist advised me: "Don't overthink it. Just be friendly, have a game plan, and want to say stuff". So I needn't be shy. I just need to write and share my work in ways that inspire you. How will I do this? I'll attempt to make the connections that others can't see. I'll find the angle: the different perspective, the unexpected view, the truth, the humour, and the revelation. I'll reveal, enlighten, inform. I'll entertain the idea until it inspires you and me. If I say things how I see them, and speak clearly, then my message will be heard. Or will it?

Finding my voice as a writer is only half of the challenge. Enabling people to discover it, to hear what I'm saying, will be the other half. Why? Because it's a noisy world out there. I'll need to find ways to differentiate myself and be heard, moving out from under the cloud of internal restrictions and preconceived ideas to find new ways of engaging my readers. Whilst books are the traditional medium, audio and video media enable easier and quicker processing of information. As a writer, should I really be locked away, writing onto a notebook or typing onto a computer keyboard? Perhaps I should be speaking my words into a camera or microphone instead? Probably. But I love writing. It's what I'm best at doing and what I most want to do. It's my terms of engagement, for now. I know that the best game plan is to 'do what you do' better than everyone else, executing in ways that have the biggest impact so that one's words register and resonate with people who recognise them as 'their thing' too. As a writer or promoter of words, one must learn the game, play the game, and then change the game. That's the mark of a leader rather than a follower. But it's not just about 'how' one delivers, because people remember the message more so than the medium. It's the message that breaks through the noise. As was taught to me by my marketing professor:

"One should always have a clear message. Why? Because confused people rarely take action". My message? It's simple: 'Stop – Unplug – Escape – Enjoy'. But this might prove complex if a broad audience interprets it in different ways.

The more complex one's message, the greater the need for human interaction. So if anything's going to hold me back, it's shyness and my insistence on penning rather than speaking my words to a room-full of people. Hiding away in my study will limit my chances of success. Eventually I'll have to get out there and meet prospective and current readers, living up to their expectations and being congruent with my message. Not much to ask of me, then? Just a Promise that must be delivered and consistently adhered to, else my brand will be ruined. Fine when talking about my worldview, but harder when reporting other things. "Trust me, I'm a journalist." How's that for an oxymoron? Better to state that I'm a non-fiction writer who sometimes prevents the truth from getting in the way of a good story. Hmm. I think I'll be having words with myself about that one. Integrity as a writer is everything. My words must always speak the truth, even if they occasionally stray into the realms of fiction. (They're there, laid out in black and white. If you want to interpret them differently, then that's fine. The grey is in the blending of ideas to form something new, which will be the case in every reader's mind. I hope you understand?)

The burden of reasoning, it is said, should be on the author and not the reader. If you have to stop and figure out my message, then I've failed. I should therefore write for you, engaging with you on your terms. The question is: do you want to read my message, or would you prefer to listen to it, watch me deliver it on film, or engage with me in *two-way* communication, perhaps face-to-face or over the telephone? Is my being a writer *enough*? Will I have to do more than put words on a page? Oh, it's so challenging. Breathe, Fennel. One word at a time. The rest will follow.

Who is my intended audience? It is those who seek a quiet, meaningful, rural life. Their heart is telling them to slow up, to seek out something different, to appreciate more, to recover, to savour a slow-paced and more cherished lifestyle where they have the time, space and energy to do and enjoy the things that inspire them. If they're like me, they'll be seeking a simple life, an outdoor life, a traditional life defined by old-fashioned customs, handcrafted things, and seasonal living. They'll be seeking things that cost nothing but which have the greatest value. They'll know that modern life is moving too fast and that industrialisation is polluting and destroying the natural world. They'll care for the environment and seek an organic life connected to the earth, which nourishes us and the soil rather than poisoning our world. They'll want to have fun, knowing that some things can get too serious for their own good. They'll be proud of their idiosyncrasies and eccentricities. Fiercely individual, they'll buck trends and be the person they want to be. They – indeed you – may have all these things and connect immediately with my vision and message, or they may feel that they need these things but don't know how to get them? If my words can help, then they'll help.

Success is not a destination, but it fuels one's journey. It pays to acknowledge and value the minutiae of success. These mini-successes remind us that we're always moving forward, even if it sometimes feels like we're stationary or moving backwards. That's why I love writing. Each word is penned to support my goal: to inspire, inform or entertain you. Each page adds to the strata of legacy. From the pure challenge of 'nothing' that comes from conquering the blank page, to the satisfaction of publishing one's work for the benefit of others, writing is the essential act that encourages me – and hopefully you – to live a meaningful life. It's the importance of today, captured in words, that commits to and secures our tomorrows. It's when the message prevails and the writing remains; when a writer's year becomes a lifetime.

IV

JOURNEY OF DISCOVERY AND RECOVERY

My mission is to be an author. I *will* keep writing in support of 'the cause'. Eventually I'll complete these Journals and move on to writing something different, perhaps humorous fantasy or screenplays for TV sitcoms. Something that makes us laugh and isn't confined by the limitations of reality. But things are going to take longer than expected, because I'm deferring my destiny. I need to prioritise other things. Actually, not 'other things' but someone else. Fate, it seems, has a good way of getting us to focus on that which matters.

It's not so long ago that I spent the best part of a year living in blissful and selfish isolation away from everything except the natural world. I turned my back on society and sought an intensely isolated existence. Now I'm yearning for the opposite. To be close to someone special. I want *and need* to be there for them. All of the time. You see, one of my closest friends is seriously ill. Struck down with bacterial meningitis, debilitated due to muscle atrophy and now unable to walk, Mike Winter has been robbed of his 'vivre' and mobility.

The man who was my fishing companion, gardening chum, correspondent, confidante, mentor and principal supporter of me becoming an author, is fighting for his life. I owe it to him to be there for him. I'm going to put my lifestyle writing on hold so that, when I'm not visiting Mike in hospital, I can write letters to him on the subject he loves most: fishing.

Like me, Mike is a keen angler who loves fishing for wild carp. He also loves the adventure and freedom of fly-fishing and the camaraderie and humour of traditional angling. That's three styles of fishing that I'll be able to write about in my letters to him – which will become chapters of future books. Whilst these subjects might appear to be an unexpected blip in an otherwise coherent lifestyle story, it's important that I write about them. I'll be doing it to inspire my friend back to health. If he's unable to escape his hospital bed, then I'll help him to escape through my letters. Given that he loves the freedom to be found in travelling light with a fly rod, I'm going fly-fishing this year. But I also want to give him some big news. So I'm going to be searching for wild carp, too. That means I'll end up writing two books this year, which will also support my commitment to becoming an author.

I'm mindful that Fennel's Journal is currently only a handwritten letter that's sent to twenty of my friends, all of whom are fishermen. Should the Journal be published, then it will be distributed to a much bigger and diverse audience. I'd expect only a handful of my future readers to be anglers. The rest might think that angling is a daft pursuit, battling boredom while waiting for a fish to bite. They're unlikely to know the adrenaline that builds between bites or how fish can swim through an angler's dreams, but these people are likely to know the appeal of sitting quietly amongst nature – allowing wildlife to come close and one's stress to roll away from one's body like mist from a mountain. Being alone amongst nature, with fresh air to breathe and new thoughts on the breeze, is an act that replenishes one's soul. It is sunlight from heaven.

Angling encourages our roots to seek out watery places and flex with the flow of life. It's a very special and seasonal lifestyle, moving from river to lake and possibly to sea, fishing for different species at different times of year. This enables the countryman to understand and synchronise his or her activities and thoughts with the natural world. It encourages a fascination with natural history and an affinity with wildlife, which often leads to passionate protection and conservation of the environment. Hence why I love angling, for its restorative qualities and ability to bring likeminded country folk together. But I don't like the way it encourages people to become obsessed with results. Anglers inevitably seek to catch more and bigger fish, which leads to competition. We have enough rivalry in our daily lives. Better for angling to counterbalance that pressure. So, whenever I write or speak of fishing, I'll be talking about *pleasure* fishing: which means taking it easy at the water's edge. This is easy for me, because I have a secret: I'm no longer fussed about catching fish. I did all my eager catching of fish when I was younger and keener. I've since mellowed and now prefer to spend my time relaxing at the waterside. Sure, I enjoy catching fish, but it's not the reason why I go fishing. I simply want to enjoy my day. But I can't tell this to Mike, at least not yet. Instead, I'll write about the thrill of the chase and the sense of adventure that comes with the pursuit of wild fish. This is more compelling to me than the end result: a fish in the net or on one's dinner plate.

I liken angling, and indeed most forms of hunting and foraging, to bushcraft and the way it encourages a deep understanding of the natural world. One reads the landscape or waterscape differently and is more at home in the wild. The more technology we can leave behind, the more we can reach out for Nature to guide us. Hence why I prefer to travel light when fishing, using simple tactics and – wherever possible – handcrafted tackle. Some of this is antique or old fashioned. But its age is less important than its organic origins. (I prefer a willow creel to a nylon bag.) The more natural the experience, the better. As is cited in bushcraft circles: "Knowledge weighs nothing". It's about experiencing real life, valuing one's discoveries and constantly learning. Which is why wild carp have appealed to me for so many years.

I began searching for wild carp twenty years ago, when I was fifteen years old. I learned that they were brought to the United Kingdom in medieval times as cultivated fish that were intended for the dinner table. But not all were eaten. Some survived and, through successive generations, reverted to a feral form akin to the original wild carp that evolved in the River Danube. We anglers call these feral fish 'wildies', although they're not true wild carp. They are, however, incredibly rare and special. Fishing for them is a form of archaeology that enables us to hold living history in our hands. And these fish aren't big like their cultivated cousins. They're small, lean and athletic. Which suits my style of fishing. With wild carp, biggest is not best – rather 'oldest is best'. The older the strain, and greater the degree of feral reversion from the cultivated form, the bigger the reward for the wild carp hunter. My ultimate challenge is to discover a strain that's survived since medieval times. Mike achieved it in Devon, finding wildies that live in Roman and Norman fishponds. Wouldn't it be great if I could continue the search, reporting progress in regular letters to Mike, and actually find the *ultimate* wild carp water? One where the fish have escaped predation or angling attention for half a millennium. I'd buoy my friend's spirits and encourage him back into health. That's what I'm going to do. I'm going fishing, for Mike.

Stop – Unplug – Escape – Enjoy

To what lengths would you go
to help a friend in need?

V

ESCAPISM AND ADVENTURE

Fishing forged my love of outdoor places, natural things, and country living. It's my foundation, my core. Having fished for more than 30 years, I've seen a lot of change in angling. Actually, I've seen a lot of change in *my* angling. Changing interests and circumstances have compelled me to fish in different ways, for different fish, in different locations and with different degrees of focus. But however I've fished, it's always been for fun. I am fundamentally a pleasure angler – always seeking to enhance and deepen my love of angling, my appreciation of natural history, and my enjoyment of the waterside.

I started fishing when I was a toddler, catching salmon parr with a net. Then I landed a trout on rod and line when I was six. I took to fly-fishing when I was eight years old. This quickly became my favourite method of fishing – especially for trout while on holiday in the Welsh hills. Whilst I've dabbled with other techniques and species over the years (I fished obsessively for carp and barbel during the nineties), fly-fishing for trout remains my favourite form of angling.

Fly fishing, which represents nearly all of my fishing these days, brings the sport back to basics, closer to nature, and to a gentle pace as it 'matches the hatch' and stops for lunch. It's one of the simplest forms of angling, closely related to the origins of the sport where an angler would cut a stick from a hedgerow, attach a line and hook and use an insect for bait. Eventually the insect became an artificial fly tied from feathers or fur, and the line became heavier and longer so that it could be flung or 'cast' to fish that were feeding further away. It's simple fishing which, in my mind, makes it a pure form of angling.

Modern life moves too quickly for me; adult responsibilities demand compromises that often force us to do things we'd rather not. I'm always looking for ways to balance these pressures by getting outdoors in timeless surroundings. Of all the possible ways to escape my troubles, fly-fishing is the most effective, leisurely, and enjoyable. I am, after all, seeking to connect to the natural 'flow' of life.

I like 'simple' and I love 'slow'. I crave being able to do things at a pace that allows me to study and appreciate them. A butterfly rarely sits still long enough for me. But I'm an artist. I like to study and 'absorb' things in detail. Fly fishing helps me to do this. Artificial flies are works of art, casting a fly is an art form, and a wild trout, salmon or grayling is a thing of beauty. Water reflects our hopes and wishes, becoming the canvas upon which our dreams are painted.

I'm at the extreme end of the artist's lens when it comes to angling. I fish for something more than fish, preferring to pursue quiet and relaxing times when fishing. I love freedom and adventure – travelling far into the quiet places that liberate me from the burden of everyday life. (I never want to be reminded that I'm only a mobile phone call away from the 'real' world.) This is why I like fly-fishing with bamboo rods and vintage tackle. They enhance the illusion of timelessness and the romance of tranquil, rural scenes. If I can convince myself that I'm fishing 'between times', then I can begin an adventure that seeks to connect with something deeper – another dimension where time doesn't matter and I'm living 'in and for' the moment. It provides continuity to my life, like pressing a pause button so that I can catch my breath and enjoy things as they've always been. I can be standing there, rod in hand, about to cast to a rising fish, and feel exactly the same as I did age six when my father first handed me a fishing rod. It must be the same emotion felt by anglers throughout time. Izaak Walton knew it in 1653. I know it in 2009. Fly-fishing energises us, helping us to live more 'compleatly'.

Just because I fish 'organically' with tackle made from natural materials, doesn't – by default – make me a traditional angler. Really I'm a rebel who likes to do things his own way. I like to buck trends, break rules and disregard doctrines. I'm passionate but not precious about the dogmas of angling. People, especially young people, need as few hurdles as possible for them to discover and enjoy the sport. So why lock fly fishing in an elitist and hundred-year-old bubble of 'upstream dry'? For something to remain exciting, it has to remain *interesting*. Which, in my experience, means that it has to be going somewhere. 'Never changing' is boring. It's rarely long before passengers depart a static train. So whilst fly-fishing is one of the oldest and purest forms of fishing, I'm not going to get locked into its ivory tower.

Today I'm on a different and unchartered path to most fly fishers, one that's seeing me blend fly-fishing and bushcraft skills to create a sort of 'modern-day primitive fishing'. It's not about spearing fish or setting nettle-spun hand-lines; it's about veering further from angling's well-trodden path to get back to basics. I want to cut my rod from a hedgerow, make my own fly-line from horsehair or nettle fibres, tie flies from whatever I can find at the waterside, make my own hooks from thorn and bone, and cook my catch over a waterside fire – just like our ancestors did. But I don't think that people are ready to read about this just yet. It's a little too radical, or retrospective, for their tastes. I'll have to take them on a gradual journey from domestic comfort to 'rewilded' adventure.

It will be several years before I share my 'bushcraft fly fishing' stories, by which time I'll most likely be fishing and foraging for my supper and living wild in a wood somewhere. So, to begin with, I'll provide a conventional view of fly-fishing. But you and I know that it won't be as new to me as I'll reveal. But hey, that's fishing: dangle a bait (or fly) that appeals. I am, after all, writing for my friend Mike who yearns for familiar pleasures.

Stop – Unplug – Escape – Enjoy

What unchartered territory,
metaphorically speaking,
would you most like to explore?

VI

TRADITIONAL ANGLING EVOLUTION

I'm about to write the letters – and ultimately the book – that I least want to write. But they're the ones I'm most expected to share. They're going to be about 'traditional angling': a form of bait fishing that prioritises old-fashioned angling values such as seasonality and sporting ethics. It's the epitome of Izaak Walton's 'contemplative recreation', avoiding competitiveness, one-upmanship and any thought that 'bigger is best'. It's pleasure fishing in the truest and purest sense.

I've long been associated with traditional anglers. My nickname 'Fennel' came from my earliest adventures with celebrated traditional angler Chris Yates; my Fennel's Priory brand name was given to me by Bernard Venables, the original 'Mr Crabtree' and elder statesman of traditional angling; and I've been 'in' with The Golden Scale Club (reputed to be the UK's top twenty-one traditional anglers) since 1996. I've been learning from and fishing with the best for nearly twenty years. Hence why I'm expected to write about the subject. But it's not something I want to discuss. Why? Because traditional angling is not what it was.

Today, traditional anglers seem to be preoccupied with using vintage and traditionally styled tackle such as bamboo rods, centrepin reels, wicker creels and quill floats. Whilst these things are lovely and enhance the aesthetic appeal of one's time beside the water, they aren't the 'be all and end all' of traditional angling. Just because someone uses vintage-styled tackle doesn't automatically make him or her a traditional angler. At least, not if his or her angling antics are urgent, competitive, ungentlemanly (or unladylike) and unseasonal.

Being connected to one's environment, moving and changing with the seasons, is a defining part of being a countryman. The traditional angler, being an old-fashioned countryman, should fish for different species in different locations using different tactics at different times of year. Traditional angling is *different*. It opposes the modern approach to fishing where the angler often specialises in one species, using one tactic on one water. (Not withstanding, of course, that an angler is likely to have a favourite specie, water, tactic or tackle item. We'll allow him or her the indulgence of favouritism.)

Writing about the type of traditional angling that I practised in the nineties, when the coarse fishing season existed on all waters and traditional tackle was harder to come by, means that I'm going to be describing something that isn't representative of traditional angling today. But I'm going to give my old school view in the hope that it persuades 'modern-day traditional' anglers that there's more to their sport than tackle and catching fish.

To avoid my writing appearing as a crusade, I'm going to use material that I wrote in the late nineties – before I became aware that things were changing. My *Traditional Angling* journal, therefore, will be a snapshot of how angling used to be. How wonderfully 'traditional' of me. First, though, I have to document my current thinking – to set the record straight before moving on, or rather back in time. So here goes:

"Fennel Hudson: the man who wrote the book on *Traditional Angling*, long-time member of The Golden Scale Club and a leading advocate of fishing traditionally." Hmm. 'Leading advocate'? This might once have been the case, but not anymore. Traditional angling's evolved, and so have I. I'm still very much a traditional angler, valuing the pleasures of fishing, the excitement of catching fish, and the heart-warming aesthetics of angling, but I'm uncomfortable with the way this style of fishing has become fixated with tackle and seeks to govern itself with a new set of 'traditional' rules.

Back in the 1990s, when I fished mostly for carp, barbel, tench, perch, roach and chub, traditional angling was the ultimate form of pleasure fishing. It sought to make the angling experience more 'compleat' by appealing to the angler's romantic side. The beauty was in the detail. From the swirling dawn mists upon a mirrored pool, to the hooting of owls at dusk, to the timeless beauty of one's attire and tackle, traditional angling lifted our spirits at every opportunity. It was passion for the heart and poetry for the soul. Gentle, calming, sensitive and diverse, it was very much a reaction to the stern-faced and ultra-competitive modern style of angling epitomised by specimen and match fishing. Sadly for me, traditional angling also became serious.

I am not a serious angler, sour-faced when the fish refuse to bite. I'm a pleasure fisherman who seeks lazy times beside the water. Rarely do I angle for fish. Mostly I fish for relaxation, connection with nature, and an escape from the fast pace of urban life. When I'm in this mood (90% of the time) I'll bring out the vintage gear and slow things down. But sometimes I seek a bigger challenge. This is when I fish larger waters that require a different approach – such as loch-style fly fishing from a boat, or long-range casting with heavy leads. Naturally, and traditionally, I use tackle that's appropriate to the challenge. Fibreglass or carbon rods can handle the exertions of casting heavy lines or leads better than their bamboo cousins, so I use them when required. They look different, so I settle in to a different style and 'image' of fishing. Does this make me a non-traditionalist? In some people's eyes, yes it does; but I'm never eager, always fishing in a leisurely manner to appreciate the nature around me.

Not everyone thinks as I do. There are those who succumb to the beauty of pretty things*. Tackle 'tarts', as they are known, hoard and caress items of fishing tackle much in the same way that red-light floozies 'collect' their lovers. Whilst the act is real, it lacks depth of connection for me. I dislike the term and the association, especially in relation to vintage tackle or traditional angling, and instead seek a more meaningful, respectful and fulfilling relationship with a waterside environment.

**The exceptions are talented craftsmen and women who make their own fishing tackle, which – given the skilled workmanship and handcrafted output – has raised their angling to a higher form of art. (Traditional angling has always appealed more to the artist than the scientist, the romantic more than the pragmatist, as it celebrates the aesthetics of angling more than the mechanics and practicality of catching fish.)*

Then there are those so-called traditionalists who are all-too serious, eager and stressed-out about their fishing. They're the ones who have 'target fish' and conduct military-style 'campaigns' upon a water (just like modern anglers). They need to catch fish, all of the time, else they've somehow failed as anglers. They conceal their desperation in tweed and canvas, hoping that no one will notice their insincerities. But 'image does not a traditional angler make'. They are modern anglers in traditional attire. Does this make them the frauds of the riverbank? Not really. They're on a journey that might ultimately see them relax and see the bigger picture. They'll eventually learn that there's more to fishing than catching fish.

Finally, there are the traditional angling snobs who create rules for other traditional anglers (I'm at risk of being one of these, given my strongly-held opinions), stating what tackle and tactics are allowed and condemning anything or anyone that falls short of their ideal. Angling has always been the great leveller of Man, so I don't seek to be a snob. Irrespective of background, profession, or social position, one should just be able to 'go fishing' and enjoy one's day – using whatever tactics or tackle that help us to relax. This is especially relevant to young anglers and those taking up the sport. Just let them go fishing and not be constrained by too many dos and don'ts.

What, then, is traditional angling? It's escapism, for sure. It's eccentric, with anglers seeking to go fishing in a way that convinces them that they've stepped out of time to a more 'organic' age. In doing so they can get closer to nature. But mostly it's a rebellion: a movement that began in the 1970s as a reaction to the cheerless face of the modern sport. Led by celebrated angler Chris Yates, traditional angling gained momentum through his articles in the angling press, many of which featured adventures with his friends in

The Golden Scale Club. He injected a great deal of eccentricity and humour into his writing and fishing, to 'balance the scales' of modern specimen angling (especially modern carp fishing) with the seemingly lost art of pleasure fishing for coarse fish. The Golden Scale Club opted to use 'vintage' split cane rods as a way of demonstrating to others that they appreciated the gentler and more beautiful side of angling. Their tackle became the emblem of their defiance to the modern sport. And then Chris proved that big fish could be caught when pleasure fishing. He caught the British record carp while stalking small fish in the shallows of Redmire Pool. A 51lb 8oz carp caught on a one-pound test curve split cane rod, 6lb line and a size 8 hook. Such light tackle, and such skill, to catch a bone-fide monster. It made the angling headlines and Chris developed a large following of anglers who sought to fish like him. But as the years passed, people forgot the origins of the movement. They also forgot that the tackle used by Chris and his friends in the 1970s was not especially old. The B James MKIV rods for example, which are synonymous with traditional angling, were made from 1952 to 1966. They were, at best, twenty years old when used by the newly formed Golden Scale Club. That's not exactly 'vintage' as we now know it. It would be the same as someone today using tackle from 1990. Sounds extreme? Younger generations of anglers are already referring to tackle and tactics of the

'90s as being 'vintage'. (Don't believe me? Check the listings on eBay.) And why not? Many of these anglers weren't born when the tackle was made.

I can imagine a young angler today saying, "Cor, wouldn't it be fun to fish like they did in the *last century*. Wow, those were classic times! John Wilson in *Go Fishing*, Matt Hayes in *Total Fishing*; we'd be seen as totally 'old school', at odds with the fashion-led modern scene. And wouldn't it be great if we caught a huge fish 'by accident'. Those trend-following anglers would never understand how we did it!"

Sounds eccentric, but when the traditional angling 'revolution' began in the 1970s, the revolutionaries were recreating a type of angling practised only twenty years earlier. And yet, here we are, thirty-five years later, still doing – and trying to master – the same old thing but with less awareness of the tongue-in-cheek humour behind it.

Anglers have 'traditionally' used the latest tackle and tactics to catch their fish. So to be a traditional angler, in the strictest sense, one *should* use the latest innovations. The so-called traditional angler, therefore, is not traditional. He or she is retro. Whatever the definition, traditional or retro angling is a reaction to the modern sport. It's laughable, though, how 'unmodern' the specimen angling scene was in the 1970s compared to present-day.

I was born in 1974 and began fishing with rod and

line in 1980. I'm a relative youngster compared to my friends in The Golden Scale Club (many of whom are now in their sixties and seventies), yet – at the grand old age of 35 – I'm finding myself feeling out of touch with the latest trends in angling.

Bedazzled and confused by the latest gizmos, I find myself looking back with rather a lot of fondness to the coarse fishing tackle and tactics of my youth – when tackle shops stocked Optonic bite alarms, monkey climber indicators, canvas brolly overwraps, and 'composite' carp rods with two-pound test curves. Yet this was exactly the sort of tackle that the originators of traditional angling thought typified the 'ugly modern scene'. Hmm. Time sure has a way of making us see things differently.

Compared to modern-day sonar fish detectors, aerial drones and underwater cameras, remote control bait boats, bite alarms synced to one's mobile phone, and electric generators that power in-bivvy televisions and beer fridges, the fishing of the 1970s, '80s, and '90s was by no means 'modern'. In fact, it seems archaic and laughable that something as simple as a braided hooklink could have been condemned as being 'unsporting' and 'untraditional'.

With modern coarse fishing being progressively technology-led, the traditionalist's reaction needs to be bigger. One rod is used instead of four, and the twenty-year-old rods that served the rebels of the

1970s have become the fifty-year-old rods that serve present-day traditionalists. It's a creaking of intent: that our gaze is ever deeper into the past as we seek to maintain the pure pleasures of angling.

But traditional angling has lost its rebellious intent. Too many people have jumped on the bandwagon and sought to rationalise or dictate the sport. There are now too many 'rules' about how a traditional angler should fish – right down to the type of fabrics to be worn or baits to be used. What began as a revolution in the 1970s has, like the punk music that existed at the time, become diluted by the masses. Actually, it's not so much 'diluted' as 'controlled'. It's become regimented, with an unhealthy amount of tackle elitism and condemnations regarding tactics. And it's entwined itself, becoming something of a 'sealed knot' for those who appreciate angling history and enjoy using vintage fishing tackle.

The problem with 'movements' and new ways of doing things is that they eventually suffer from the same oppressive dogmas that caused the revolution in the first place. Traditional angling is no different. Holding it in a pseudo time warp of tackle imagery that never appeared together at the time (Edwardian clothing, with 1950s rods and 21st century hooks and lines) makes for many grey areas of what is acceptable and what is not. And these areas are the most contested. It's started a fight within a fight, which might cause the ultimate undoing of the movement.

Anglers are a cliquey bunch. Their conflicting interests, enthusiasms and passions (which verge on religiousness) encourage love-hate relationships. There are a great many rifts between those who fish in specific ways or for dedicated species. Consider the styles and attitudes of game, sea, coarse, pleasure, specimen, match, traditional, modern, and 'carp' angling (which many anglers, wrongly in my opinion, regard as a dirty four-letter word). Such contrasting imagery, stereotypes, rules, mindsets and doctrines. For angling's survival, anglers should come together to defend and support the sport. Alas, the conflict caused between the factions means it is often a case of 'never the twain should meet'.

It's not often said, but the 'rules' of traditional angling are built upon the highly combustible shavings of bamboo rods mixed with cravings for a 'golden age' that never existed – at least not as it's imagined today. This sounds brutal, but halting one's sense of nostalgia is as limiting as trying to prevent progress in angling. Why should someone in their twenties or thirties be required to dress, tackle up, and act as someone from the 1950s? Rose-tinted re-enactment aside, they should be allowed to feel nostalgic for whatever's happening during their own lifetime.

I'm about to say something that I've shied away from sharing for two decades. It is this: the higher you climb in traditional angling circles, the more intensely passionate and purist the fishing becomes.

The bright flame of inspiration and achievement burns brightly. There's great beauty in this, but also ugliness. It can be an elitist, pretentious, condescending, cliquey, ego-driven, dogmatic and political scene that favours the few and condemns the many. At its finest it's a side-splittingly madcap bastion of hope for everything that is cherished in angling. At its worst, it's a self-serving corruption of what is otherwise a beautiful sport.

After nearly twenty years of fishing traditionally with fellow traditional anglers, I've concluded that the upper echelons of traditional angling are – in places – the opposite of what I imagined them to be when I first picked up a split cane rod. My fishing diaries from the 1990s are filled with 'confessions for my sins' because I was made to feel so inadequate by those whom I respected. I laugh at these comments now, but in my formative years my peers criticised me heavily for doing things 'incorrectly'. For example: I was chastised for using an umbrella when fishing in the rain. Doing so, apparently, was too close to becoming a 'bivvy boy'. Instead, I was told to use a rubberised canvas poncho and huddle under it during a downpour. Later I was ridiculed for using fixed spool reels. "Can't you cast thirty yards with a centrepin?" they laughed. "What an amateur traditionalist!" And then the final insult: caught red-handed using Heron bite alarms. They were forty years old and previously owned by their inventor Richard Walker, but they used *electricity* – something absolutely forbidden by the purists. I was again hauled up in front of The Golden Scale Club and forced to apologise for the ultimate sin: "using alarming bite sindicators!" My punishment? I would have to float lighted tea candles down the River Test at Stockbridge. If a trout rose to eat one, and the flame wasn't extinguished, then I would be reprieved. But until then I would have to live in shame of my deed and scurry about in the gutter with the rest of the wannabee traditionalists. Sounds funny? I didn't see any humour in it at the time. It destroyed my confidence, soured my love of traditional angling, and ended some previously close friendships. Hence why I temporarily distanced myself from The Golden Scale Club. (An organisation that denounces others with 'awards' such as the Black Gaff, Black Pyke, and Black B*llock seems rather self-righteous to me.) But I still love traditional angling, especially the gentle style practised by our forefathers, and my closest friends are still within The Golden Scale Club. Traditional angling can still be pure, if you maintain the right mindset (knowing that it's nothing to do with tackle) and realise that there's more to fishing than catching fish. It's a recreation, not a religion. Keep it in perspective.

Those who belittle others, or seek to prevent change, restrict innovation. Like museum curators, they preserve the past instead of welcoming the present. This, in its place, is good. But so is creativity and adventure into new things. (So I should stop harping on, and accept that traditional angling has evolved.) Embrace the richness and fullness of angling. Is a cork-handled and expertly whipped carbon rod so wholly terrible? Is a modern reel that runs smoothly, has a bail arm and clutch that works, worse than its clunky fifty-year-old grandparent? Who is to say? And who is to judge? Think differently about what is traditional and what is modern. (Clue: everything will become vintage; so buy quality items now and make them last.)

Keeping traditional angling locked in a pseudo 1950s trance is not healthy. It needs to change, mature and evolve. It needs an influx of something new. What is right for one is not necessarily right for the other. That's what makes one's uniqueness and self-confidence so appealing. When it comes to fishing tackle, use the best of what is old combined with the best of what is new (so long as it looks right and feels good together). If you buy and use quality things, they'll never go out of fashion. Use the right tool for the job. Split cane rods were not designed to cast long distances with heavy leads or lines; centrepin reels are not designed to cast over forty yards; wicker baskets will dimple your buttocks if you sit on them for more than an hour; a sleeping bag laid directly on the ground will soon become damp and uncomfortable; and night fishing without a bite alarm will lead to sleep deprivation and illness. There's a reason why modern tackle items were invented. Use them. Who knows, they might become traditional some day?

Prequel to The Quiet Fields

VII

IT'S IN OUR NATURE

I'm a countryman. I was born and raised in the country. Most at home when I'm outdoors, I'm happiest when I'm amongst nature. I'm a gardener, woodsman, naturalist, angler, bushcraft practitioner, and environmentalist. I'm also a writer, living in fear of the person I once was.

Back in 2003 I wrote a wedding speech that intended to solidify my way of thinking. All it did was highlight how misguided I was. I never got to say the speech, so I had a lucky escape from an alternative fate, but I wonder whether the letters (and subsequent Journals) I've shared since then have unintentionally communicated my real way of thinking? Of course they have. Fennel's Journal is one big manifesto about living a good-humoured, appreciative, nature-based, meaningful life. It's not a fishing story.

Spending three years writing exclusively about angling has proved to me that, as much as I love the sport, it plays but a small part in my overall appreciation of country life. My angling writing helped to inspire Mike Winter during his illness, but his health is still deteriorating. I need to write with more passion and deep-rooted enthusiasm if I am to buoy his spirits further and continue on my path to an 'authentic' life. I need to get the narrative of Fennel's Journal back on track by writing about things that *really* interest me.

As an all-round countryman, my writing is very much in the vein of the naturalist-sportsman writer 'BB'. With this year marking the 30th Anniversary of his book *The Quiet Fields,* I have decided to take a rest from writing about fishing and pay homage to BB by theming my letters around countryside subjects. I want them to read more like my earlier writing, perhaps as follow-ons to *A Waterside Year*, to form a new foundation for what may follow. If Jack Hargreaves could build his message bit by bit with his *Out of Town* TV series, then I can do the same with Fennel's Journal. It's about living a holistic life, with everything in balance. So I'm going out and about, seeking to get back to what's within. This year will see my writing 'rebooted', which requires me to don my wellies and introduce you to a new world: a quiet place, a wild place; where we may savour things that really matter.

The Quiet Fields are where we can take time out from the stresses of modern life, to stop the wheels for a while, unplug from the daily grind, escape to an idyllic and peaceful place, and enjoy the natural world. They're where we can connect with nature, study wildlife, savour rural lifestyle, and understand how the 'truly great' outdoors helps to shape our identity and passions. They're where we can seek and find freedom and adventure, reflection and contentment. Ultimately, they're where we can achieve a slow-paced and meaningful life.

Whether we live in the countryside or not, The Quiet Fields are where we can savour the joys of country life. They're where we can hear birdsong, see blossom on a hedgerow, and feel something deep within. They're where we can celebrate the seasons, feel the natural pulse of life, and talk about the joys and challenges of living and being in the countryside. And, as we're part of nature, I'm talking about the nature that's within us as much as around us. But for us to travel there together, you need to know a little more about me. After all, we're here *together*. Just you and I. Awaiting the dawn.

So, who am I? I'm Fennel, the writer reborn. As of this year, I'm introducing myself as a 'rural lifestyle & countryside author'. This means that I'll be writing about nature and the outdoors, how travels in wild places enable us to explore the notions of freedom and self, and how a slow-paced existence helps us to appreciate and reflect upon the subtle and precious gifts of life. I'll observe the small things, the gentle things, the significant things that otherwise might go unnoticed. These things of beauty, however small, are the building blocks of a fulfilled life. I'll seek a meaningful and contented country life that's lived on my terms, not succumbing to the beat of someone else's drum. I'll believe in individuality, personal identity and life purpose: that we should always know and remain true to ourselves. And because of this, I'll always encourage you to "Never do anything that offends your soul". (Sounds like a big 'ask'? Not really. I don't ask for much. I'm a simple man who likes nothing more than sitting quietly amongst nature, where stillness within encourages much movement and life around me.) I'll seek freedom in the 'quiet corners' of the landscape that allow my soul to breathe.

Energised by the natural world, I'll be most at home in the countryside. That's my authenticity, my calling. And it should be. The countryside is where I was born and raised; it's where I live and work. If my boots cease being muddy, then I've walked too far along a well-worn path. That's when I veer away, into the undergrowth of the unknown, to find comfort in the familiar. Why? Because the countryside, to me, is home. It's where I'm most happy, where I can proudly say that I'm a *contented countryman*. I'm also someone who's seeking to connect with you, to find common ground – quite literally – in the countryside. I'll speak, you'll speak, we'll speak. We'll speak to friends. We'll listen. And we'll hear. Because it's about sharing the message: that life in the woods and fields is so vibrant, so energising, so 'contenting'.

But these are early days in my writing. It's New Year's Day 2011; I'm still living in the quiet recesses of the countryside, writing these words for you to read in the future. I'm the one who can spend all day sitting beneath a tree, motionless but for the movement of my pen across a notebook on my knee. I'm existing quietly, not seeking to make a noise with my voice. At least not yet. Soon I'll be speaking up and speaking out. It's inevitable. My friends are urging me to share my writing with others. My message, it seems, is too loud for just twenty people. They're asking me to copy my letters and send them to others, perhaps even posting them to a website where anyone, anywhere, can read them. This doesn't seem like a quiet activity, so I'm struggling to know how to speak loudly while remaining quiet. I don't want to disturb the nature within and around me. I want to remain a quiet man of the woods and fields. But I'm committed to becoming an author, so I'll need to overcome my lifelong shyness. Ironically, to overcome my fear of others, I'll need their help. I'll need them to help me spread the word, they'll need to twist my arm and kick me up the backside to get out in front of people, and they'll need to guide me on what they'd like me to write. And I'll need to be absolutely committed to my beliefs, so that the pain of not doing something is greater than the fear of doing it. My message *has* to be shared.

The message in my writing is: 'Stop – Unplug – Escape – Enjoy'. But the 'stop' bit doesn't mean for us to – how shall we say? – 'cease'. It means for us to stop doing the things that prevent us from living. Ultimately, I seek to be a contented countryman. To be contented, one has to be *free*. Freedom is what we seek by venturing into green open places. But is it a place, or state of mind? Is contentment something to be found outside, or within? If something makes us contented, then it's helped us to move from a state of *dis*contentment. Which means that something else must have been making us feel like we needed something more, something *different*. It's born of an 'unease' that things aren't right. That something is out of balance, or lacking, or wrong. Some people might say that it's in our nature to be discontented, that – through eternal restlessness, materialistic cravings and ambition – we always want more. It's how we progress, how we *evolve*. Stop. Listen to your intuition. It's a powerful thing. It gnaws at our conscience. Listen to the 'gut feel' that tells you, from the depths of your soul, whether your life is in balance. (If it's not, then do something about it. Never allow yourself to get dragged into the mire of 'if only'.)

Balance, for those in control of their destiny, is achieved through the choices they make. As Jerome K. Jerome wrote in his 1897 book *The Idle Thoughts of an Idle Fellow*: "It is impossible to enjoy idling thoroughly unless one has plenty of work to do. There is no fun in doing nothing when you have nothing to do". 'Balance', therefore, is best achieved when we're pulled in both directions, to savour the pleasures and pain of both extremes. As I wrote in 2006: "You have to experience the bite of the wind to appreciate the warmth of a winter coat".

For me, being outdoors in green places is more than necessary. It's *essential.* The longer I'm away from these life-giving places, the less I live and the more isolated and 'disconnected' I feel. Like an ape in a zoo, I gaze through the bars of my urban making and yearn to be free. No wonder, therefore, that the *Oxford English Dictionary* describes 'contentment' as: "a state of happiness and satisfaction. For example: he found contentment in living a simple life in the country". Isn't it great that the definition of contentment involves a simple life in the country? We know it; they know it. It's official. The countryside brings contentment. Such is the power of The Quiet Fields.

Where does this idealised vision of contentment come from? The green and pleasant land, rose-tinted and filled with country-folk sitting on hay bales watching the sun go down? My view is that it's from literature. Great and compelling writing from poets, natural history authors and storytellers. Romantics like William Wordsworth, with his *Lyrical Ballads*, and Thomas Hardy with his *Far from the Madding Crowd.* They call to us, beckoning us so seek the rural idyll. A better life, away from the madness of the modern world. But a country life is a tough one; it's defined by practicalities and hardships. It's about working the land in all weathers.

It can be brutal and relentlessly laborious. But so can urban living. 'Nine-to-five?' It doesn't exist anymore. We're 'always on, always connected'. But connected to what? White noise? No wonder the silent beauty of a star-filled sky is such a celestial treat.

There's great wonder to be found in the natural world – if we make time to stop and observe it. Then, when we finally see, we can study, reflect and learn. As Charles Dickens wrote: "It is not easy to walk alone in the country without musing upon something". This, for me, is where the very best naturalists exist. The great nature writers such as Gilbert White, Richard Jefferies, John Clare, Henry Williamson, and W.H. Hudson, are all known for writing in the 'rural tradition'. They see and reflect, making observations that we can compare to our existence. My personal favourites are the sporting countryman 'BB', who wrote classics such as *The Idle Countryman* and *The Wayfaring Tree*, and the rural lifestyle author C. Henry Warren. They knew that the countryside is a living, breathing, evolving thing. They didn't overly romanticise 'the good old days', but instead sought to ensure that traditions continued into the future. Which is exactly what my writing will seek to communicate.

I remember C. Henry Warren recounting in his 1945 book *England is a Village* how one villager said to him: "They say times are better now, with less work and more pay. They say we worked so hard and so long that we hadn't any time left to enjoy ourselves. I dunno so much about that. Why, in them days, after ten or twelve hours' scything in the harvest fields, you'd hear men go home at night singing; and it weren't always for the beer inside 'em either!" Isn't it strange how this image of contentment goes 'hand in hand' with outdoor labour? That physical activity in the outdoors can provide contentment?

Hence the term 'contented countryman'. 'Countryman' means someone who lives or works in the country. The 'man' part comes from the Latin word 'manus' meaning 'hand'. To be a countryman, therefore, is to be a 'country hand'. Which is why we don't say 'countryperson', however politically correct it might seem. Male or female, a countryman is someone with his or her hand in the country – be it work or play. If their heart is there, then being there (in thought or body) provides their contentment.

The Quiet Fields – open places, green places, fresh air and a sense of freedom – energise us, lift our spirits, make us happy, healthy, and contented. Treasure these places. As 'BB' wrote in his 1950 book *Letters from Compton Deverell*: "There are thousands of people like you, and me, who wish for nothing better than to enjoy these delights while yet we may…Only be thankful, that is the main thing, only be grateful for this gift of life." Wise words from the author whose books include the inscription:

"The wonder of the world
The beauty and the power,
The shapes of things,
Their colours, lights and shades,
These I saw.
Look ye also while life lasts."

So, 'look ye also'. Seek and find The Quiet Fields. Be grateful for the gift of life, appreciate the countryside, and – above all else – strive to be contented.

Prequel to Fine Things

VIII

THE FINEST OF THINGS

What would happen if you woke one morning and didn't recognise yourself in the mirror? If you couldn't remember your name or age, or what defines you? What if the only clues to your identity were the things around you? What if they could somehow communicate *who* you are, and *why* you are? What if they could be your way of finding your way back to who you *were*, and in turn who you *will be*? And, hope of hopes, what if these things – and the stories contained within them – could help you *smile*? I woke to a predicament such as this in May 2004. I'd existed in a sorry state of denial and depression for six months following the loss of everything I'd previously held dear. The trauma had erased my recent memory. I had to look around the room to find reminders of my being. There was a birthday card from my parents – the people who'd been there for me no matter what and whom I loved completely – and my childhood bedroom furnished with things from a time 'before'. How long before? From before I'd morphed into the person I'd become, but was no longer. I knew this. I hadn't just wiped clean six months of memory, but several *years* of foolish behaviour. I was reborn, renewed, refreshed, and repurposed. I began to remember the loss – of things taken, and things sold to pay my creditors – and the desire to hold on to the things I most valued. The big things had gone, yet the small things remained. Why had I held on to, and fought to keep, some things and not others?

Here's the list of what I was left with after losing everything else I'd accumulated in thirty years: an old pair of woollen walking socks, a cotton shirt, a pair of moleskin breeks, a pair of walking boots, a flat cap, a red waistcoat, a wax jacket, a cheap pocket watch, a bundle of letters contained within a burgundy box file, a tea chest full of old notepads and diaries, an ordnance survey map, two bottles of seventeen-year-old writing ink, three photo albums, two dozen books, four boxes of 'stuff' from art college, a rusty pruning knife, an old fishing rod, four fishing reels, some cork-bodied fishing floats, and a box of deer hair fishing flies. Oh, and three bailiffs orders, a final notice from a credit card company, a mortgage foreclosure, a letter from my solicitor, and a bank statement showing that I had more debt than could possibly have been amassed by a sane person. Yet I hadn't lost everything. I still had the love of my family and this collection of 'things'.

I got out of bed, dressed myself in the reassuringly comfortable clothes, then sat down and began reading through the box of letters and crate of notebooks.

They helped me to remember who I was and what's important to me. Themes appeared, around self-belief and self-confidence, freedom, adventure, wild places, natural history, countryside traditions and rural living, outdoor hobbies such as fishing and gardening, and a whole realisation about truth, honesty, integrity, passion, beliefs and striving for a balanced life lived on one's terms. Above all else, the notebooks conveyed my passion for writing. This, apparently, was the act that most defined and inspired me. I'd signed them all 'Fennel'. Other than the legal letters, bank statements, and the birthday card from my parents, there was no mention of 'Nigel'. It seemed that he was lost with all the other possessions that had disappeared. Instead, I'd chosen to keep 'Fennel': the person I was before and would become again. All thanks to the items, which I named Fine Things, that served as the breadcrumbs to my former self.

It's now 2012. I've spent eight years rebuilding my life and becoming an honest representation of the person I was before. I've collected many Fine Things since then, all of which remind me of who I am. So I'm going to write about these things, hoping that people will appreciate the importance of sentimental items and places. But this year will be different. No longer will Fennel's Journal be a handwritten letter. It's going to be a high-quality magazine, a nod of respect to my friend and mentor Bernard Venables.

He did it with *Creel* magazine in 1964, so I'll do it with Fennel's Journal this year. I promise that it will be a very Fine Thing.

However, in secret, I'm nervous about the authenticity and integrity of my lifestyle. Putting things into print seems like an official endorsement of my message, but amongst all the excitement of professional publishing, I have a growing concern that I might be an echo of the past rather than a child of the future. My terms of reference were fine for my past self, but are they appropriate for today? I've retraced the breadcrumbs of myself back to 1996, when I was twenty-two-years-old and when Fennel's Priory began. It's when I was happiest: living in a rented one-bedroom cottage on a country estate in Berkshire, working as a gardener, spending every evening going for walks through the woods, then writing letters into the early hours; I'd fish at weekends and crave nothing more from life. It was the happiest of times. Or was it? I was single, lonely, struggling to pay my bills, and unsure about my future. Surely I'm happier now? I became a father last year; I have a caring wife; a house of my own; more 'things' than ever before; closer friends, better bonds to my family; more interests; and more happiness than I've felt in a long time. Perhaps, for the sake of forgiveness and gratitude, I should separate my present state from what came before?

My current happiness is no longer dependent on

how I lived sixteen years ago. I have a new set of rules and a new foundation for the future. My finest things are my daughter and wife, and the times we have together. Materialistic things could never compare to them in value or sentiment. But I'm not going to write a magazine about my family. There's a limit to how much of my private life I'll share with others. So I'm going to pay tribute to the things that helped me to rediscover myself and the actions that reveal my personality. That way, I can read the stories as my way of remembering 'me'. I'll not hold anything back. I'll give access to my impulsive and eccentric mind, revealing the mania that's come to define the good times. I hope the readers of my magazine will laugh, I hope they howl; I hope they frown and ponder; I hope they study the things around them and identify their own Fine Things. Most of all, I hope they – and Mike – are entertained by things that were *painfully serious* to me but which now are laughably madcap. Its message will be: "Know thyself. Know thy things. Know when to laugh and when to cry. Know what you value, and why". Oh, and know when loo roll and a bottle of ginger beer turns me into 'that guy'. And to ensure I don't censor the content, I'm going to write, design and publish the magazine in nine days. That's how long I have before the copy deadline at the printers. So I'd better get writing…

Stop – Unplug – Escape – Enjoy

Is your 'self' formed by your past, or built for your future?

IX

THE ROOTS OF LIFE

What's within? The intuitive calling; our DNA, our heritage, our ancestors living within our bones, our spiritual resilience that – if we're true to our roots – keeps us alive. One's sense of who we are goes way beyond the things we own, or the activities that define us. It resonates to our core. It's from where we came. Our ancestry, ancient and recent.

I'm hugely indebted to the wonderful upbringing gifted to me by my parents. They are the ones who imparted upon me their love of the outdoors – particularly gardening, fishing, bird watching, hiking, and camping. They also taught me what is right and wrong when it comes to one's behaviours and values, and one's responsibility to the natural world.

Very quickly I learned about the organic cycle of life. My paternal grandfather died two years before I was born, so he wasn't able to meet me – his first grandson. But I got to know him through the fishing tackle and books he left to my father and in turn to me. Fishing, from a very early age, was my way of filling a very noticeable void in my family's life. One's grief, I discovered, turns not to sadness but to hope when someone's legacy is evidently alive in others. Every time I went fishing, I was perpetuating the memory of my grandfather. But, biologically, he is only a quarter of me. The dominant part of my make-up, which closely supports my being an angler, is that of being a countryman. Why? Because my mom's side of the family were farmers. They'd farmed the land around Bridgnorth in Shropshire for more than four hundred years. This long bloodline of people connected to the land gives me my greatest sense of belonging. It's no wonder I became a gardener and why I seek a self-sufficient life, rooted to the land. Farming is in my blood. I am connected to the earth, the rhythm of the seasons, and the life that grows from nothing. 'What was, still is', so long as I trust my instincts and stay true to my ancestors within. This reminds me that the organic cycle of life is very much like the act of gardening – where we sow seeds and nurture them into maturity until they finally return to the soil.

The fragility and finite nature of life is topical at present. I'm mindful that my friend's health is failing. Mike has contracted cancer, so his prospects don't look good. 2013 might be my last chance to cheer him up with some funny stories and remind him that I'm still striving for a better life. I'm going to pursue the idea of a 'gardener's year', where gardening is a metaphor for allowing things – and one's dreams – to grow.

There's always hope. Even when we've returned to the soil, we're enabling other things to grow. This lifts our heads and hearts, knowing that death and decay feed the roots of future life.

Let me be clear: *A Gardener's Year* will only be superficially about gardening. It's likely to be a metaphor for the organic cycle of life; the life-giving 'nature' of the land; and the fun, hopes and dreams that are to be nurtured when living one's life to the full. It's going to be about connection, to one's origins and to the natural world.

Scottish mountaineer W.H. Murray wrote: "The heart's natural movement is to lift upward, and this is most readily done in the wild, for there it is easy to be still". I'm sitting still at the moment, observing a sunrise through my kitchen window. It's early spring here in the UK. The first snowdrops and celandines are blooming; cherry blossom and hazel catkins will soon be showing in the woods; hawthorn leaves will appear on sun-drenched hedgerows and some early daffodils will brighten the short days. Already I can hear the songs of blackbirds, robins, wrens, tits and finches. It's as real as we can get. Imagining this makes me feel *vibrantly* alive.

My gift, passed down to me by my parents, is that of countryside wisdom born of observation and caring. I know that the deepest pleasures lie in natural things. I feel the yearning when, at the peak of a busy day, I crave to be 'away from it all'. But what is 'it'? The 'it' that

I seek to be away from? Surely I don't want to be away from it *all*? At least the 'all' that matters. Don't I mean that I need to be 'close' to it all – insomuch as my 'all' is amongst the life-giving wonder of nature? Perhaps, when I'm starved of my 'fix' in the countryside, I crave fulfilment? I need 'topping up' and plugging in to an altogether more natural current – the current of life. Away from it all? No. I prefer to say, 'Away from The Nothing, and close to it all': connected to the nature within and around me.

Nature connection, in my mind, is more than a reaction to the artificial 'always on' virtual world of the Internet and modern 'connected' living. I love the countryside for its own merits. If ever I'm in doubt of this, I refer back to my favourite countryside books – the ones written in the early part and middle of the last century, when mobile phones, computers and televisions were unknown. The authors spoke of the countryside with great affection, because they knew of its special importance.

One of my favourite quotes is by W.H. Hudson, from his *Afoot in England*. It says: "The charm of the unknown, the infinitely greater pleasure in discovering the interesting things for ourselves than informing ourselves of them by reading. It is like the difference in flavour in wild fruits and all wild meats found and gathered by our own hands in wild places and that of the same prepared and put on the table for us.

The ever-varying aspects of nature, of earth and sea and cloud, are a perpetual joy to the artist, who waits and watches for their appearance, who knows that sun and atmosphere have for him revelations without end…[but] he knows that his striving is in vain – that his weak hands and earthy pigments cannot reproduce these effects or express his feeling…he has his joy none the less; it is in the pursuit and in the dream of capturing something illusive, mysterious, and inexpressibly beautiful".

Hudson's words were written in 1909. That's 104 years ago. Yet his message is perhaps more pertinent now than when he wrote it. The study and appreciation of natural history is about getting out there in the wild, to observe nature first hand. It's where, as my namesake said, we can experience life, to connect with nature and know the 'inexpressibly beautiful' qualities of the natural world. It's Nature's garden. The very best there is.

As with everything in nature, it starts small. So I will begin my 'gardener's year' – defined by nature connectedness and the realisation of dreams – by doing the simplest of things: opening a window. I'll look and listen, feel the breeze, smell the scents, imagine being out there cultivating the soil and my future. Soon I'll discover a world of natural things delighting my senses. The journey, therefore, starts close to home. We don't need to look far. But when we can, we should gaze and breathe deeply. As the naturalist John Muir wrote:

"We are now in the mountains and they are in us, kindling enthusiasm, making every nerve quiver, filling every pore and cell of us". This, in its entirety, is what I call 'Nature Connection'. It's about reconnecting – to the roots of life. That's what I'll seek to achieve this year, as I pen *A Gardener's Year.*

Prequel to The Lighter Side

X

REACHING FOR DREAMS

This prequel is made up of two parts. The first was written at the start of 2014: the year I completed my ten-year plan to rebuild my life. It was when I put an end to all the sacrifices, compromises and slavish hard grafting that had enabled me to clear my debts and have enough left over to purchase a dilapidated but 'picture perfect' cottage in the Cotswolds. It was to be a new beginning for my family and me; a brighter, lighter side of the mountain we'd climbed. But it didn't go to plan, hence why the second part reflects upon the 'lighter side' learning from this unexpected year.

Part One – Of Geese and Grass

The golden goose…is not golden. It looks like any other goose, just that it lays golden eggs. So while you kneel there, arms outstretched, with hands cupped awaiting your prize, you realise...what? That the thing in front of you still poops, honks and hisses like any other goose.

Waiting for the golden egg is a messy business. It's not a nice experience, nor does it come from a nice bird. But you tolerate the goose. Why? Because it lays golden eggs. At least, eggs with a thin gold shell.

Seeing the egg appear from a goose's bum. Eww... Cleaning off the muck with your shirtsleeve, to see your reflection upon the gleaming surface? Hmm. Are we seeking the prize, or how it makes us look? Better to walk away from the goose, and the egg.

The grass, our colleagues and friends will caution, is not necessarily greener on the other side. But it can be. It can be lush and cool and fresh and invigorating. "But it still needs mowing," they'll say. We know that. We enjoy the prospect. But they fear it.

Who, exactly, is 'they'? They're the ones who dream of better, but fear change. 'Different' scares them, so they lie low, accepting the norm. They wait.

We – the committed ones – strive for better. We seek the best. We dream of being. We seek happiness. We're compelled to find things we can't see. We hope and believe in The Promise. We know the truth: that burying our hopes is not a trade for money. Sooner or later we'll feel the need for change. For something better and more rewarding. We won't wait. We'll make.

There's joy in seeking and dreaming. It's the wanting that makes us take action. Because, deep down, we know what we want: to break the egg, to walk upon the grass, to get to what we seek. It's our reward, for working, saving, striving and dreaming...of this – for *ten years*. Now's our time. Embrace the moment. We deserve it. Dreams *do* come true.

Part Two – Survival

2014 arrived and with it the expected completion of my ten-year plan to rebuild my life. I paid off the last of my debts from 2004, moved to a cottage in the country, celebrated my 40th birthday and – promptly – had a breakdown. It was infinitely worse than the one I'd had in 2003, magnified by the irony that it was caused by my efforts to build a better life. The relief of finally reaching my goal, mixed with deferred exhaustion from the climb, proved too much. I collapsed, grieving for the life I wanted but didn't have the strength to enjoy. And I mourned the loss of Mike Winter, who had passed away six months earlier. This could have ended my dream, and Fennel's Journal. But unlike my illness ten years earlier, I decided not to mope but bounce back with more passion than ever. I kept going – knowing that fulfilment was very nearly within my grasp.

I remained focused – and conscious – remembering the importance of using one's energy to do the things we love. Which, for me, means writing. Lots of it. So I took six months off work to explore the lighter side of life. And then I wrote about it. *Passionately.* The result is a deeply felt and wisely-considered Journal that's the natural companion to *A Meaningful Life*. It's a collection of lifestyle essays that addresses key issues such as work-life pressures, the importance of faith, keeping going through adversity, how to see the funny side of a situation, and – bold move that it is – a study of the depression that's so common in creative people. It's sounds heavy, but it's not. Fennel's Journal is resolutely upbeat and optimistic, marked by positivity and awareness that reading is a form of entertainment. It's about the *lighter* side, so I hope you enjoy it.

Hmm. That last paragraph sounds like marketing spiel. This is a book of secrets, so what about the stuff I *don't* talk about in *The Lighter Side*? About how one's pursuit for the ideal is skewed by individual perspective, preference, and interpretation – something that's difficult to process when one suffers from high and low moods. What might be ideal for one mood could be catastrophic for the other. That aside, I'd overlooked how shamelessly selfish I'd been in rebuilding my life.

I've spent ten years recreating my old-fashioned, isolated, self-sufficient rural life from 1996, doing my best to become the person I was back then. It's right for me, but is it suitable for my family? Mrs H and Little Lady were bought in to my vision, but they'd never experienced this life before. I'm the only one to have lived so remotely as to cherish the absence of neighbours, shops, road traffic or streetlights; and the lack of technological intrusions such as television, radio, Internet and telephones. They, on the other hand, have felt desperately vulnerable being so cut off. A cottage in the middle of nowhere is great when you're alone. But when you wake with a fright in the middle of the night, hearing someone trying to kick down your front door and smash in your windows, it takes on a viciously unforgiving face. This was our fate from the first week we moved to our cottage, and has been so ever since, even though the police are patrolling past our cottage twice per day. With violent and spiteful neighbours intent on forcing us out of the area, and with me working overseas for weeks on end, Mrs H and Little Lady are literally 'holed up', petrified by whatever will happen next. How could I have been so selfish as to impose such a wretched existence upon them? I'd so desperately wanted a perfect life in the country, and I'm sure we could have achieved it in the right place, but I am responsible for imposing a living hell on the people I love most.

My advice? If something looks perfect, it's probably not. We've found the Cotswolds to be a 'film set veneer' concealing a far-from-perfect backdrop. Our life here is terrifying, stressful, intimidating, awkward and impossible, made infinitely worse by me refusing to accept that something I've worked so hard to attain is so completely wrong for my family.

Mrs H is very different to me. She likes her coffee shops and high street fashions, her cinemas and restaurants, her modern things. Sure, she loves 'quiet'; but she's more suited to modern, urban living. I've only recently learnt this. I have to respect it, and whatever tastes Little Lady adopts. But how do I live with the compromise? I still yearn for a little cottage in the country, miles from anywhere, with more garden than floor space and more wildlife than furniture. I *crave* simple. I yearn for a quiet life. A self-sufficient life, with livestock and orchards and vegetables. Away from the death-knell of a chaotic, high-stress modern lifestyle. Or do I? Is this the 'want' of my former self? Am I now too old to labour this dream? Would I not prefer central heating, double-glazing, a low maintenance garden, local shops and a hospital close enough to resuscitate me when I eventually keel over? Would I not prefer a house devoid of damp that doesn't crinkle my books and ruin my clothes? Or one free from mice that scurry over one's duvet at night? Would I not prefer to live in the town?

Whilst a warm, modern house appeals, I would not enjoy living in a town or city. I've done this before and found the traffic noise to be unbearable and close proximity of 'overlooking' nosy neighbours to be intrusive. But I'm not averse to meeting my family in the middle, perhaps living in a relatively modern house if it's built with character and is located in a quiet spot on the edge of a village. A barn conversion, perhaps, or a new-build timber framed cottage? That would be an acceptable compromise, so long as it had a garden big enough to feed my family – and keep the neighbours away.

Why so much focus on 'home'? It's only a building.

As is said: 'A house is made of bricks and beams, a home is made of love and dreams'. It should be possible to be happy anywhere. And besides, I'm an outdoorsman, why should I worry? Surely my focus is on getting 'out there', away from buildings and people? Don't I just need somewhere that's right for my family and merely a stopover place for me? Where I could sleep before venturing off into the wild during the day? No. I'm a homely chap. I need to be at home with my family. I've spent the past ten years travelling all over the world, doing whatever's required to clear my debts. It's time for me to come home. I want this more than anything. I'm also a gardener. I'm unfulfilled unless I'm in my garden, working the soil and growing things. But I'm not sure that my family wants a self-sufficient life. (They might tolerate chickens, but pigs and sheep are out.) And as much as it pains me to accept it, I'm not sure they want me around all the time. They need their space as much as I do.

The purpose of 'home' is something I hadn't adequately foreseen. The thing I'd missed in my 2004 plan was how the people central to that plan – Mrs H and me – would change during the ten years it would take for the plan to come to fruition. Perhaps that's the big flaw in a plan based on rebuilding a former life? That one's lifestyle, interests and health change as one gets older? I forgot to factor review points and adaptability into my plan. I've been so busy rebuilding the past that I've forgotten to accommodate the future.

The danger in reaching for one's dreams is in striving for the wrong dream. It's too easy to focus the 'thing' that manifests one's achievements, such as a house, rather than the lifestyle that surrounds it. No point slaving all our lives to buy our dream home, to then not have the energy, time, or knowhow to enjoy it. One's lifestyle is more important than one's possessions. Doing what inspires us, at every opportunity, is what gives us strength for the next adventure. It's far better, therefore, to have a very clear picture of what we'll *do* – rather than have – during our journey. Because, so long as we're alive, the journey never ends. There is no 'one day it will all be perfect', because if you haven't got your perspective sorted right now, you'll only be taking a load of negative baggage with you into your dream. A skewed view will prevent you from accepting that 'good enough' is good enough and that your destination will not be marked by perfection – rather contentment with what you've got.

Being with my family defines happiness for me. I value their company more than anything else. I wish to be with them and care for them every day. Without them I am lost. That's the one and only wilderness I fear: the sensation that I feel lonely when I'm alone.

I'm going to make the required changes, accepting that I got it wrong. I'll renovate this cottage and sell it as quickly as possible. We'll move somewhere new. I'll get a new job, closer to home. We'll begin a new and better life. We *will* get through this.

Everyone is programmed to survive, so we should trust our instincts. Do what we know to be right. Do it with the people we love.

Stop – Unplug – Escape – Enjoy

What's your life-defining dream?

XI

FRIENDSHIP MATTERS

I'm going to publish a book that proves that things can be achieved when true friends pull together. Why? Because I've spent the past six years fighting to create such a thing with the wrong group of friends. More specifically, with a group of friends that includes two people who apparently have forgotten what friendship means. Sounds contentious? It is. But I'm going to tell you the story all the same.

Back in May 2009, a group of friends assembled in a pub garden in Dorset for their Annual General Meeting. Only it wasn't a *formal* meeting, rather a jovial get-together. But this gathering was more sombre than in previous years. One of the members was noticeably absent, struck down with a near-fatal illness. He'd been in a coma, was battling muscle atrophy and skin ulcers, had apparently lost the use of his legs, and was only slowly regaining his mental faculty.

We needed our friend back. Back amongst us, back to health, and back on form. This would take time. His consultant stated that it might take up two years for him to regain his strength, so he would remain under specialist care for the immediate future. Eventually he would return home, but only when he had the necessary welfare facilities installed – including an adjustable hospital-style bed with crane to lift him in an out, and a wheelchair-friendly shower room. This would cost money. Money that wasn't readily available.

I informed the group that our friend's first words, when waking from his coma, were "A Shower of Golden Scales". Knowing that he wasn't the Terry Thomas type, we figured that he wasn't referring to the group as an 'utter shower'. Rather, he'd imagined a book written by the group. It would be entitled *A Shower of Golden Scales.*

I proposed that the group should write the book, and that each member – past and present – should be invited to write a chapter. They could write about anything they liked, so long as it captured the interests and character of the group. It therefore wouldn't be a book *about* the group, rather a book *by* the group.

We agreed to produce the book – for our friend – and donate all royalties to a fund that could contribute towards the cost of bringing him home and providing him with a better quality of life.

I agreed to co-ordinate and edit the contributions; the group photographer would provide the photographs; the secretary would provide access to the archives; and our well connected 'fixers mate' would secure a publisher. Everything went to plan, with 68 chapters submitted,

172 photographs provided, 104 handwritten notes scanned, and a publisher waiting eagerly to produce the book. It would be a quality publication, available in vellum- leather- and cloth-bound editions. There was even the prospect of doing a 'Club Edition' bound in fish scales. It was exciting, to the point where our friend was encouraged so much by its progress that he was helping me to review the drafts of the book. It was a lot of work, but worth it for our friend.

Alas, all the activity came to nothing.

Sadly, two members of the group were slow to submit their chapters. The book wouldn't be complete without their contributions and blessing. And then they got twitchy, concerned that the book might contradict the 'secret' nature of the group or not represent the quality of writing they'd expect. They raised their issues: "Was taking a 'bottom up' approach from the group the right thing to do? Would it create a collective character for the book that might be different to what might otherwise be achieved if it were written by a select few?" Tensions began to mount. Then the unthinkable: rumours began to circulate that the real reason these individuals were holding back was because they thought the book might compete with their soon-to-be-published titles. Was it true? I doubt it. But it alarmed several members of the group, who pulled out of the project. Momentum drained away and I was left trying to explain to our friend why things were taking so long. Sadly, he died in 2013 and all momentum for the book was lost.

Fortunately for our friend, I'd foreseen the concerns of the few and had hatched a contingency plan. Back in 2012, when I launched Fennel's Journal as a magazine, I sent him the first copy. He and I talked at length about what the Journal could become. It had captured his imagination, much as 'the Club book' had done three years earlier. He enquired whether I'd consider including guest chapters within the Journal. I agreed, testing the model with three chapters in the magazine version of *Wild Carp*. It worked, so I devised a plan to create an edition of the Journal written almost entirely by guest contributors. The theme would be that of friendship, something I was passionate about after seeing how true friends could pull together for a worthy cause. Of course, our friend was the first to submit chapters for consideration. They must have taken him ages to write, as his mind was often muddled and his hands were weak, but the quality of his writing was good. I invited others to submit chapters. They responded – and delivered. I quickly amassed enough material to publish a special edition of the Journal. So that's what I'm going to do.

This special publication will be entitled *Friendship* and will contain stories about one's awareness of time and the wisdom of age. It perfectly captures the bond of 'old friends', demonstrating that Fennel's Journal is a publication 'for us and by us' with friendship as the theme. It proves that friendship really does matter, and that true friends will make the effort.

2018 Postscript:
I'm so grateful for the friendships that have arisen since I began writing and sharing Fennel's Journal. They've taught me that friendship is given freely. There doesn't need to be a reason or a request, as the company of friends enhances our lives. We're there for each other. Simple as that. And we're honest enough to say 'thank you'. This was proved by the magazine edition of Friendship, with so many friends freely contributing material for the magazine. It was also the most successful edition, selling out in less than twenty-four hours. I invested all profits, including those budgeted for the book edition, in a celebratory meal for the contributors. I hosted it at The Flyfishers' Club in London. The toast, as you would expect, was "To Friendship".

As for A Shower of Golden Scales*? I doubt if the book will ever receive its two missing chapters or be made available for sale. There's only one copy in existence. It sits on my bookshelf as a reminder of what could have been, of a farce that should never have been, and of friendships that are no more. The title is perfect. We really were an utter shower.*

Stop – Unplug – Escape – Enjoy

What could you and I do
to enhance our friendship?

Prequel to Nature Escape

XII

MAKING THE ESCAPE

We've done it! Mrs H, Little Lady and I have escaped our two-years of hell in the Cotswolds. We worked every spare minute of every day and night to complete the renovations to our cottage and get it fit for sale. We slept on the floor, in amongst dust and rubble; we cancelled our holidays and social plans for a year; we taught ourselves how to plumb, plaster, lay bricks, fix woodwork, glaze windows, replace roof tiles, and ignore the vermin; we endured two winters with no adequate central heating (the warmest we could get our house, last winter, was 6 degrees centigrade – so cold that I couldn't bend my fingers to type or write). And staying up decorating through the night discouraged the local yobs from further vandalising our house. Sure, they took their opportunity while we were away at work, throwing buckets of urine and faeces at the house, covering our gravel driveway in nails, snapping trees and pulling up plants in the front garden, and posting leaflets through our door about how we were 'too young to have earned the right to live in such a place', but while we were at home we were not bothered – and not afraid. And we stayed clear of the pub, so there were no further attempts to beat me up, hurt Little Lady, or intimidate us with verbal abuse. The house went on the market in April and sold within eight days. The new buyers were Russians 'with connections' and thus had no concerns regarding neighbourly disputes. They'd secured their perfect Cotswold cottage. Full asking price. No questions asked other than "How quickly could we complete?" We were out of there, car tyres screeching, on our way to a new life in North Wales (to be close to Mrs H's family) by July. We never looked back. The ordeal was over.

The Promise, to which I committed to live my life in 2006, said: *"Life can be impossibly tough. At times it seems like there is no escape from either the pressure or monotony of a world where we are corralled into being something or someone we are not. But there is an escape, to more peaceful and gentler places, and a happier state of mind. It requires a conscious decision to view the world in a certain light, searching for and spending time at these places. Not everyone can pursue this dream to the full. My promise to you is that I will do everything I can to live this life as completely and happily as possible, and will share the journey with you."*

Our time in the Cotswolds proved that sometimes, for one's safety and sanity, a physical escape is required. There's only so much the 'certain light' (positive outlook) can do to make good of a bad situation. But The Promise was absolute, which meant that I would never accept a fate that fell short of the ideal. I would keep striving for more peaceful, gentler, and happier times. For me, my family and you.

The Fennel's Priory mantra is made up of four words: Stop – Unplug – Escape – Enjoy. It's an easy to remember, step-by-step route map to a better way of living. Sounds easy? Not always, and certainly not for me. Not everyone knows how to stop doing the things that distract them from what they love, or that these things are a distraction at all. Many people struggle to unplug from the 'always on' connectivity of mobile phones and Internet-enabled devices that suck them into an unreal world, or escape the shackles that lock them into a captive indoor existence, or enjoy freedom when they find it. They know that they want something different or better, but something is preventing them from achieving it. Almost always, there's only one thing stopping them: themselves. Their mindset and conditioned behaviour is locked into the system. It's a system governed by fear of change, anxiety that 'different' might end up being 'worse'. So they keep going, trudging towards the unseen enemy while carrying a dead donkey on their back.

Eventually these people learn that the best way to survive No Man's Land is not to trudge around with Eeyore on their back, dodging bullets as they go, but to lie low and make themselves small. If they can hide out of sight, then maybe their problems will go away?

Ultimately they will, often when the person reaches retirement age and lacks the strength to do the things he or she wanted to do when younger. But they've escaped the ordeal and can now enjoy the quiet life – if that's what they want and if they can still remember how to enjoy it. Please, I beg of you, do not end up in this situation. Do not live passively as though your life is not yours to live. Be bold. Live *actively*. Take ownership. Not just of your future, but of your mind and the way you interpret things.

Act, don't react. Make your escape, if you need it. But would you be escaping *to* or *from* something? Stay focused on the better life that awaits you – aim for 'to' not 'from'. I've run from recent problems in the Cotswolds, but my focus and energy has been on getting my family to safety – to a place where we can relax and soak up the beauty of somewhere we can genuinely and heart-meltingly call home. That's where I am now, sitting in my garden looking out to the appropriately named Hope Mountain. I'm living on the Welsh border, between the Clwydian Mountains to the west and the Cheshire Plains to the east. Borderland country, just two miles from the River Dee and only a hundred yards from woodland and farm fields; yet just a short drive to Chester and school for Mrs H and Little Lady. It's the best of both worlds, here on the border, with one wellie-clad foot in Wales for me and the other (in polished leather stiletto) pointing towards England for my girls. We're here, happily rooted to place and people, enjoying our new home and rediscovering what it means to be free from stress, debt and oppression.

I'm going to celebrate my 'escape' to North Wales by capturing the pleasure of 'natural freedom' in a book. I'm going to call it *Nature Escape*. It will highlight how wonderful it can be to escape for the pure pleasure of what we can enjoy when we're there. There will be no running away from anything, just an open-armed desire to make the most of a special place.

With Mrs H and Little Lady due to be away at Nana and Grandad's this weekend, I have an opportunity to spend twenty-four hours doing whatever I want. The thing I most desire, which I missed while I was in the Cotswolds, is time spent writing in a woodland. If I can go there, to sit quietly with a notebook on my knee, then I might be able to write about all the things that pass by me, or through my mind. It will be my way of celebrating 'freedom': honouring The Promise I made a decade ago.

So instead of writing a 'compare and contrast' lifestyle Journal, as I've done in the past, I've decided to produce something that really champions the message of 'Stop – Unplug – Escape – Enjoy'. I'll grab my rucksack, some provisions and a writing pad, and walk a hundred yards from home to my new Priory Wood, to spend a day and night alone amongst nature. I will walk, camp, study wildlife, reflect, and feel grateful for my blessings.

The result will be a Journal that provides a first hand, real-time, account of a day and night 'enjoying' – as BB would describe it – 'the wonder of the world'. It will be a book to sit nicely alongside *A Waterside Year* and *The Quiet Fields* – making a trilogy within the Fennel's Journal series, designed to help you 'escape' when you read it. So even if you can't spare twenty-four hours to sit alone in a wood, you will be able to do so by reading the book.

If you're one of the people who wonder how to Stop – Unplug – Escape – Enjoy, then I encourage you to do it and not think it. If you're reading Fennel's Journal as instructed, in a quiet place where you can savour the beauty of nature, then you've already taken the four steps needed to get there. Congratulations. You're in control. Enjoy your freedom. You have escaped.

XIII

THE SECRETS WE KEEP

Fennel's Journal, at least the main part of the story, is complete. Last year's *Nature Escape* marked the end of my quest to rebuild my life. But there are more books planned. Why? Because the series was shaped by a mistake.

Four years ago, I commissioned a luxurious leather binder to house the Journal magazines. I was very specific in my request: it was to house all fourteen Journals in a way that they wouldn't fall out when the binder was removed from a bookshelf. Once they were in, they were in for good. The binders were produced and shipped to customers across Europe. And then I realised my mistake. At the time of shipping, there were *thirteen* Journals planned: one for each year of the ten-year journey, plus three 'specials' (*Wild Carp*, *Friendship*, and a *Subscribers'* edition). This wouldn't do. I couldn't have a binder with a gap at the back. Fennel's Journal had to be *complete*. So I did the obvious: I added another Journal. And in the spirit of appearing in control, I announced that I'd always intended to publish fourteen Journals. "There's a secret one," I said, "that completes the series". It sounded convincing. But I hadn't the foggiest idea of what the 'extra' magazine would be about. All I had was a 'secrets' theme and a long list of subscribers wondering if they'd have to pay more for the additional magazine.

Crisis encourages creativity, especially when we have to act quickly. So I've decided not to overthink things. I'm going to use 'secrets' as the angle for the missing magazine, publishing a retrospective Journal that will link all the others together. This is opportunistic and necessary. I've become uncomfortable with the magazine format, as it's encouraged distinct 'theming' of standalone subjects rather than an obvious end-to-end journey. Readers are questioning why the story centres on my life and why the Journal doesn't always include writing from multiple authors as in other magazines. I need something that joins everything together and makes it clear why I themed each title as I did. So, what began as a mistake has become a blessing.

Fennel's Journal *is* a series. It's meant to be read from beginning to end. It's about a life rebuilt. It's my life, rediscovered, reimagined, and reshaped on my terms. Ten years, from 2006 to 2016. But Fennel's Journal is not a diary; it's a communication between friends. It has *purpose*. It seeks to mean something to you, so that you can use it to help shape your thinking and actions, to live a more meaningful life. That's its big secret, which isn't really a secret at all. Hence why the first title in the series is called *A Meaningful Life*. The clue was there, right from the start.

Where and when will the series end? Will it end at fourteen? The *Subscribers'* magazine will be merged into other titles when the series is published as books, so we still need a fourteenth book. I'll name it *The Pursuit of Life*, because – above all else – writing Fennel's Journal has encouraged me to live. It's made me feel alive and pushed me to do things I once feared. When I think back to the dark times of 2003, and how I wanted everything to end, I was so fortunate in being able to 'write myself to happiness'. Each word has moved me further towards a better place. So *Life* will be a book that adds extra detail to the story, capturing how and why rural living and the countryside have been my lifelines. But will it be the last book in the series?

As with that incomplete binder, something tells me it's wrong to end the series at fourteen books. Surely fifteen would be a 'rounder' number? But should I tag one on the end, or write a prequel? What would interest you? What came after, or before, the series? Perhaps it should just be a collection of stories taken from the many letters that haven't yet been published? No. I think it would be more rewarding to write something new. Perhaps a super-collectable book that's only provided as a gift for those who've most supported me. Perhaps those who've arranged a Fennel's Priory event, or done something amazing that's helped me to promote my message? That would please the Friends, but drive book collectors crazy. Could I be so naughty? I think I could. But you'll have to keep it secret, at least for now.

I've kept many secrets while writing the series, weaving in lifestyle messages as metaphors that are evident in some Journals but not others. So that there's no confusion, here are the main messages from each book in the series:

- *A Meaningful Life* defines the values, character, mantra and messaging of Fennel's Priory and the terms upon which I will honour The Promise.
- *A Waterside Year* is about recovery and self-rediscovery, celebrating the life-giving qualities of water.
- *A Writer's Year* is the book I most wanted to write, as it cements my commitment to becoming an author. It pays homage to the writer's craft but also seeks to increase the reader's self-confidence.
- *Wild Carp* is about 'the mist of believing': the importance of pursuing one's dreams – that they can be achieved, however impossible the odds.
- *Fly Fishing* is about adventure, getting close to nature, and finding freedom in wild places. It's about 'getting away from it all'.
- *Traditional Angling* explains that that angling is a recreational treat, to be done when everything else in life is settled.

- *The Quiet Fields* is about the fragility, beauty and importance of natural things – which includes us – and the joy of the seasons.
- *Fine Things* is about identity, humour, eccentricity, and how sentimental objects can contain and communicate our personality.
- *A Gardener's Year* is about the organic cycle of life, the joy of growing things and cultivating one's dreams.
- *The Lighter Side* seeks light after sadness, even if the light doesn't at first want to shine. It's about the good times.
- *Friendship* proves that great things can be achieved when true friends pull together.
- *Nature Escape* concludes the ten-year story. In it I'm escaping for a good reason, because everything in my life is settled.
- *Book of Secrets* provides the 'behind the scenes' story, revealing why each Journal is themed as it is.
- *The Pursuit of Life* illustrates how rewarding – and sometimes challenging – it's been to 'Stop – Unplug – Escape – Enjoy'.

There might be one more book, which will be given rather than purchased. And then Fennel's Journal will end. The spotlight will then no longer be on me. I'll return as The Contented Countryman, celebrating the stories and achievements of other country folk and the wildlife and quiet places that enrich our lives. But that's several years away yet. First I must finish what I've started, writing a Book of Secrets that provides the whole agonising truth of a life rebuilt.

XIV

MOONLIGHT PERSPECTIVE

I'm writing this at 3am on 4th November 2017. I'm leaning against a tree, looking out across a moonlit lake. Normally I'd be asleep at this time, tucked up in bed and dreaming of the day to come. But I'm wide awake and doing my best to stay warm as I gaze out across a mirrored pool. The moon is full and sky clear, so I can see perfectly well to write.

You might think that I'm continuing my Nature Escape, but I'm not. At least not as completely as I did. This moonlight perspective provides the clarity I need to share some unexpected developments with you. So, friend to friend, here's an update on some big news that has challenged me but ultimately enhanced my worldview and increased my writing output.

I spent ten years writing about my quest for a meaningful rural life, completing the task in 2016 by quitting corporate life, moving to North Wales and writing *Nature Escape*. I've since lived 'completely' as a family man and countryside author. I've been *immensely* happy, taking Little Lady to school, going for lunches with Mrs H, rebuilding the Fennel's Priory website, launching a podcast and YouTube channel, blogging every week, publishing three books, and spending all my free time outdoors. Alas, I was a little too preoccupied with enjoying myself to worry about whether my lifestyle was sustainable.

Sales of my books covered their marketing and production costs but were insufficient for me to draw a wage. So I lived off my savings, planning to add money into the kitty when my royalties increased. I was just about to send the press release for *A Writer's Year* when Mrs H took me to one side and showed me our bank statement. Ouch. In fact, 'ouch times ten'. We had enough money in our account to fund one month's rent. No more. After that, we would be in trouble. Something had to be done. Panic-stricken and dazed by my abrupt departure from Creativeland, I realised that I would have to do the one thing I swore I'd never do: go back to corporate employment.

Returning to the 'suited world' was something I never expected. I genuinely thought I'd 'got out' and had definitely found freedom through my writing. But family obligations must come first. So I made calls to recruitment friends and influencers, informing them that 'Nigel Hudson' – the writer with a gift for helping companies make oodles of cash – was back on the market. The 'other guy' who goes by the 'other name' was coming out of retirement. I'd be available to the highest bidder, 'selling out' to provide for my family.

All, and only, for money.

Job opportunities came in from Canada, US, Ireland and Norway. But I didn't want to work overseas again. Little Lady had only just got to know me, Mrs H and I were back being a proper couple, and I was deeply in love with the green mountains, blue rivers and pure air of Wales. I needed a UK-based job.

Sadly, the North Wales job market doesn't fare well on the salary front. So I accepted fate knowing that I would have to search throughout the UK. Finally, I received three 'considerable' offers. All of them were 'down south'. Two in London and one in Dorset. I accepted the one with the greener surroundings and better fishing. Dorset it would be. But Mrs H and Little Lady wouldn't be coming with me. This would be a lone journey, made with the sole intention of sending money home to my family.

My new job would be 250 miles from home. With favourable traffic it would take me six hours to drive there, longer if I attempted to use public transport. I wasn't prepared to commute that far every day, so I'd drive down to Dorset in the early hours of Monday morning and return home late on Friday evening.

As I signed the employment contract, I felt mixed emotions of anxiety for the role and despair that I'd be losing everything I hold precious. Daddy would be going away again, the prospect of a quiet family life in the hills was over and, it seemed, I had failed at being a full-time author. I trod the darkest path I've experienced in recent years, questioning the entire Fennel's Journal story. Indeed, 'Fennel' would have to fight for consciousness as the 'other guy' took over. Never happy bedfellows, the personality clashes would happen all over again. Like Gollum and Sméagol in their cave, Nigel would endure the darkness while Fennel searched for the light. Or would they? Perhaps I just needed to view things from a different perspective.

Reality is a matter of perspective. This I believe, as do countless writers, photographers, artists, scientists, and cognitive behavioural therapists. We choose our reality, at least how we perceive it. Life can be good or bad, beautiful or ugly, tranquil or stressful, depending on how we interpret things. We *choose* to be happy or sad. 'We perceive therefore we are.'

Mrs H and Little Lady's lives would mostly be unaffected by my being away. They were used to me not being at home. Their routine and house-proud tidiness would be better without my creative chaos and mess, so they'd be more settled without me. And as long as I sent home a regular pay cheque, then I'd still have a home and wife waiting for me when I made it back to Wales. So, in short, I would miss them more than they would miss me. My time away, therefore, would be emotionally crippling but liberating at the same time. If I maintained an optimistic gaze, I could convince myself that I was being set free. I would have all week to do whatever I wanted each night. All those hours to write, relax, and take in the natural world.

Where, though, would I live? Where would I write? Could I set up a Fennel's Priory 'Field HQ' so that I could maintain my creative output irrespective of circumstance?

A writer can write anywhere. I could write in a car, at work, or wherever I found myself to be. But where one resides can influence one's sense of grounding. Being away from loved ones isn't nice.

It was the prospect of returning to a 'hotel suitcase' lifestyle that most upset me. I'd travelled the world almost non-stop in recent years, so there was no way I wanted to see another business hotel. I'm an outdoorsman who won't be cooped up with the complimentary tea bags, Corby trouser press and miniature bottles of shampoo. I can do so much better than that. I am, after all, the person who encourages us to 'never do anything that offends your soul'. If I were going to do this, I'd do it on my terms.

So, on 5th June, I loaded up my car and headed off to face my new future. My new home would be a Nordic tipi that I would pitch somewhere in the New Forest. I'd use it as my basecamp, from where I would 'strike out' to go to work each day and explore the heathland and ponds of the forest each evening. Mine would be a life of extremes, living tramp-like at night and as smart executive by day. Plus, I'd enjoy seeing the campers' expressions as I appeared in 'Mr Ben' style from my tipi each morning. I'd have gone in dressed in my scruffiest tweeds and come out dressed in my posh suit and shiny shoes.

How did it go? 'Ups and downs' if I'm honest. The first week was a boggy washout, with gale-force winds and rain that turned the campsite into a lagoon. In fact, most weeks have seen me battle heavy rain and then load a soaking wet tipi into my car before returning home. But being so close to nature has been wonderful, lifting my spirits to counteract the homesickness that I've inevitably felt.

Highlights have been the nightjars and owls calling each evening, and the display of shooting stars in August (when I lay on the ground with just my head sticking out of the tipi, looking up at the meteorites above). My biggest nature shock, however, was having two badgers nudge under the walls of the tipi during the middle of the night and then struggle to escape. (I needed a strong cup of tea and change of pyjamas after that incident.) But the best thing that's come of my outdoor lifestyle is that I've spent each evening writing by candlelight. My weekly blogs have been written there, either in pen or on my laptop computer. (They were later uploaded to the web using the free Wi-Fi service in the local coffee shop. God bless the Friday morning espresso.) So, with the exception of my 'nine to five' office time, I've been living a very simple rural existence. It's reminiscent of my 'Waterside Year' in 2005, but unlikely to end so quickly.

After 97 nights outdoors, I'm perfectly in tune with the practicalities of outdoor living – and the lack of an iron for my shirt. I remember to hang my clothes up off the floor, where they won't get soaked if floodwater comes in; I peg down the tipi on all points and have a stock of candles for light and warmth; and I know which foods will keep without refrigeration (and which chip shops to visit when I can't be bothered to cook – usually on Tuesdays, Wednesdays and Thursdays). My new colleagues think I'm bonkers, but I'm proud to be doing things my way. I've enjoyed so much freedom and peace. I've been alone and always lonely, but I've been so productive. It's been my best-ever writer's year, even though it's cost me the close bond I'd forged with my family.

Whist forest life is lovely, it isn't as compelling as waterside life (at least water that's next to a wood). So, at the end of October (when the campsite closed) I moved my pitch to a copse that fringes a lake in the New Forest. I then went home and collected my fishing tackle (and noted how much Little Lady had grown), returning two days later to erect a smaller tent beside the lake. This gave me another new perspective, and more things to do.

Camping and fishing for four nights each week has enabled me to get closer to my new environment.

I've discovered the kingfisher's hole where it's rearing a late chick; I've watched the long-tailed tits grow and flutter about the willows; I've identified three different types of bats (which the bat detector app on a friend's phone informs me are pipistrelles, daubentons and brandts); I've seen a near-white heron and a rat the size of a Rottweiler. And, of course, I'm piecing together the movements and feeding habits of the fish that inhabit the lake.

I'm enjoying this unwanted yet surprisingly delightful existence. It's a new life, a writing life, done – as always – on my terms. Where will it lead? I don't know. So long as I pay my bills and publish my books, then my time away will be worth it. One step, one word, one night, at a time. So everyone can benefit.

PREQUEL TO THE ONE YOU CAN'T BUY

XV

NOT FOR SALE

It's true. There is a secret fifteenth Journal. It's not publicised and is only provided as a gift in return for supporting me. The secret book, entitled *The One You Can't Buy* is my way of saying thank you. It contains a very open message about the importance of staying true to one's values, identity, friends, and goals; and includes stories of how and where I've done this for our benefit. But before you read it, you should read this prequel. It explains the ultimate learning from my years writing Fennel's Journal: that the beginning was the end, and that the end seems to be the beginning.

My worldview was formed as the result of a bad decision I made in 1998, when I compromised my values in return for a conventional lifestyle that pleased my bank manager but put me on a temporary road to ruin. I also kowtowed to peer pressure, burying talents that would have threatened the egos of some so-called friends. It's taken me twenty years to realise that I should believe in myself, follow my own rules, use my talents, and do things as passionately and energetically as possible. I shouldn't be swayed by others, however well meant or manipulative their guidance or pressures might be. I'm my own man, so I'll do things on my terms. I will not alter my course for the wrong cause, person or employer. I'm a man of values and integrity. It's all or nothing for me. I am not for sale.

A question for you, given that much of Fennel's Journal was written long before it was published: is the running order of the series correct? Is the numbering of the books, and the date of the chapters, accurate? Is it chronological or should the series be read in a different order? Perhaps it's a circle where the final book leads directly into the first?

In whatever order you read the Journals, or how you interpret them, my message is the same: keep going, no matter what, but always make time to appreciate the journey. My biggest learning in writing about my story has been 'dogged determination to keep going'. Perseverance as a writer and dreamer is *everything*.

I've been writing as Fennel for 22 years. But I've not fully achieved my dreams. I'm at least five years away from seeing my vision for Fennel's Priory mature, and ten years away from it becoming something tangible (perhaps a little holiday cottage with a nice garden that can be enjoyed by others). That's more than thirty years for Fennel's Priory to be properly 'out there'. If I'd known this at the beginning, would I have started? Of course I would, just as I keep going now. I'm in it for the long game, with the ambition of creating a legacy that enhances the lives of others.

Living a meaningful life is not about short-term happiness, or about the numbers that enable success to be measured. It's about doing something consistently well so that people value the output. You've read this book almost to the end, so you must value what I'm saying. For that, I thank you. I trust you are working towards and realising dreams of your own, making a difference by being different, not doing anything that offends your soul, and always remembering to 'Stop – Unplug – Escape – Enjoy'.

Stop – Unplug – Escape – Enjoy

What could you do to receive
The One You Can't Buy?

XVI

ALWAYS FOR YOU

Here we are: *Book of Secrets* is complete. It links all titles of Fennel's Journal together. That's fourteen books (actually fifteen…shhh), documenting a turnaround in happiness, freedom, health and good fortune; where nature, the countryside and rural living are the heroes of the story. The journey will keep on going, but this stage is complete. I should end it with a fitting conclusion.

There's a greater man than me who can sum up our journey, a mountaineer who in 1865 first climbed the Matterhorn. Edward Whymper, over to you: "There have been joys too great to be described in words, and there have been griefs upon which I have not dared to dwell, and with these in mind I say, climb if you will, but remember that courage and strength are naught without prudence, and that a momentary negligence may destroy the happiness of a lifetime. Do nothing in haste, look well to each step, and from the beginning think what may be the end." That's been the underlying message throughout Fennel's Journal. It's been about real life, not perfect life, seeking out the meaningful things it has to offer. These gain value when we appreciate their context. Life isn't always good. On occasions, it's heartbreakingly cruel. But, like seedlings that grow in a wood when another tree falls, we seek out the light when it comes. Unlike the seed that sits and waits, we can go in search of that which brings us to life. We can choose to get up, stand tall, and change our future. It, like the happiness and meaningfulness in one's life, is for one's making. We choose where we are, where we're going, and how we see the world.

The key to one's freedom and success? Have an idea and act upon it. Believe in your idea, and yourself, and give it your all. Invest all your passion, conviction, courage, skill, persistence, and resources. Truly believe in your ability and cause. Get it moving at pace, and see it through to the end. It's about execution: making it happen and being restless if things don't develop as quickly as they should. Make time, though, to appreciate the journey. Find your balance. Indulge your senses by smelling the roses. Breathe deeply as you travel, but always take big and bold strides – to make and be the difference you seek.

Albert Einstein said, "To know and not to do is not to know". Know what you seek to do – and do it. In other words, have an idea and act upon it. You can plan it out, like I did, or you can just get going and work it out along the way. As a scaffolder would say, "Build as you build". Don't spend all your time dreaming and planning. You have to start sculpting. Marvel at the thing you've created, not that which supports it. It's about the end goal, your finest creation, and how it affects you and enhances others. It's your art, which you share proudly with the world.

Be strong. Those who lack courage to realise their dreams are but souls drifting upon a windless ocean. Don't drift, or sink. *Swim*. You never truly know what's over the horizon, but that doesn't matter. It draws you there all the same. It's the promises of 'different' and 'better' that keep us travelling, building and dreaming. And then, when opportunity comes, you'll find fellow travellers on the same journey. You travel and build together. Ultimately you meet someone to whom you'll pass your baton of dreams. This person is the child of your creation, embodying all your hopes for the future.

Little Lady, my books are for you. Learn from them. When I am no longer around, and you need advice, encouragement, a loving hug, or to feel safe, you'll be able to find me right here. I'll always be here, welcoming you with open arms and my biggest smile, excitedly awaiting our next adventure. I am your daddy, your best friend, the one who loves you no matter what, who's proud of everything you are and will become, who wants you to be you. Be everything you can be. Be happy. Be safe. Be fulfilled. Know that you are loved. Go in search of your dreams, knowing that you will never be alone. In our special places, and within these pages, I'll be there for you. Always.

ABOUT THE AUTHOR

FENNEL HUDSON

"Author, artist, naturalist and countryman. His is a lifestyle to inspire the most bricked-up townie."

Fennel Hudson is a rural lifestyle & countryside author known for his *Fennel's Journal* books and *Contented Countryman* podcasts. He's also a keeper of secrets, only revealing things when he's ready. His stories always have extra meaning, usually hidden within metaphor, as he's someone seeking to understand the relationship between human nature and the nature around us. At home in the woods and fields, he's an outdoorsman seeking the 'complete picture' of a balanced rural life. From his humble beginnings as a gardener, to his achievements as a writer and public speaker, he's consistently urged us to 'Stop – Unplug – Escape – Enjoy'.

For more information please visit:
www.fennelspriory.com

THE FENNEL'S JOURNAL SERIES

THE FIRST-EVER REVIEWS OF FENNEL'S JOURNAL:

"Fennel's Journal began as a series of illustrated letters to friends. As these evolved they became less a diary, more a manifesto, and the Journal is now exactly that – a way of living, rurally and simply: very real for all those who recognise the importance of tradition and joy."

Caught by the River

"I can see where it might lead. What he has would make amazing TV. It's the Good Life, but in a realistic way. It's Jack Hargreaves. It's Countryfile. It's quality Sunday newspaper stuff. It's 1948, all over again. In trying to escape the present he's inevitably created a brand. A potentially very powerful brand."

Bob Roberts Online

"Fennel's Journal is a masterpiece about rural living. It is a route-map to the life we all seek."

The Traditional Fisherman's Forum

From A Meaningful Life:

"Life is the most beautiful and rewarding gift. We just need to take time out to allow us to reflect, change perspective, and see things in their best light. Sometimes we just have to stop and feel the pulse of the Earth, the rhythm of the seasons and the internal voice that was once our childhood friend. As the natural world grows smaller, so too does its intensity and the size of the window through which it may be viewed."

NO.1

A MEANINGFUL LIFE

A Meaningful Life is the first and perhaps most important Journal. It documents the origins of Fennel's Priory and why Fennel decided to live by a new set of ideals. With themes ranging from escapism, adventure, work-life balance, identity and purpose, through to traditionalism and country living, it sets the scene for future editions – building messages that are central to Fennel's Priory. Ultimately it conveys the importance of a relaxed, balanced, and meaningful life.

READER TESTIMONIALS

"I loved reading this Journal. It's inspiring and has the beginnings of something very special."

"Fennel's chosen trajectory is firmly in the slow lane. He's a countryman, with courage to stand behind his traditional values."

"Witty and emotive, Fennel's writing conveys passion for a slower-paced and quieter life."

From A Waterside Year:

"Water is intrinsically linked to the mystery and excitement of discovering new worlds. Of dreams. And hopes. And thoughts of what 'could be'. Dreams free us from normality. ...As the daydreams grew longer, the distinction between what was real and what was imaginary grew less. Soon I existed in a blissful world of my own creation. Reality, as I learned, is only a matter of perception...A life that is real to one is surreal to another."

NO. 2

A WATERSIDE YEAR

In *A Waterside Year*, Fennel takes time out to live beside a lake in rural England. Here he appreciates the healing qualities of water, studies the wildlife around him, lives at the pace of someone outside of normal daily life, and discovers the freedom that's found in isolation. Getting so close to Nature, and spending time in idle fashion, enables him to discover a stronger sense of self. Ultimately he learns that freedom is not a place, but something that exists within us.

READER TESTIMONIALS

"A year in the wild. How we would all love to follow in Fennel's stead and indulge our dreams, to come out the other side a stronger and wiser person."

"A Journal with a message – that we should take time out to think about what's important, and see the beauty of the world."

"A truly blissful read full of inspiration and humour. The story of Fennel sitting in his tent, with the noises outside, had me laughing out loud!"

From A Writer's Year:

"Writing, with a fountain pen and ink from a bottle, is the simplest of things. Yet it can transport us to a different place entirely. Imagination is the real magic that exists in this world. Look inwards, to see outwards. And capture it in writing."

NO. 3

A WRITER'S YEAR

A Writer's Year celebrates the writer's craft. It champions the handwritten letter, discusses vintage pens and writing ink, and celebrates things such as antique typewriters and the quirkiness of the creative mind. It's a blend of observations. It's funny. It's serious. It's real life. But most of all it is written to encourage aspiring authors to find their voice, to put pen to paper, and follow their dreams.

READER TESTIMONIALS

"Worth it for the first chapter alone. It cannot fail to motivate and inspire the would-be author."

"What Fennel has written is not so much a eulogy for the handwritten letter as a call-to-arms for everyone to follow their dreams and make the most of their God-given talents. This is a genuinely inspiring read."

"I loved the part: 'If a pen can communicate our thoughts, dreams and emotions and be the voice of our soul, then ink is the medium that carries the message'. It shows how important and generous writing can be."

From Wild Carp:

"Some will say that searching for your dreams is like looking for unicorns in an emerald forest. They will say that following a golden thread will lead only to a king, dethroned and living in the gutter. This may be so. But the king was made, not born. The crown was never his to wear. ...If ever the adventure proves tiring, or you lose sight of your dream, look to the west at sunset. There, on days when the skies are clear, you might see upon the horizon a thin layer of golden mist. When it appears, you will know its purpose: it is the mist of believing."

NO. 4

WILD CARP

Angling for wild carp is about adventure, history, atmosphere and emotion. *Wild Carp* captures this aplenty, describing Fennel's 20-year quest to find a very special type of fish. But it's also about nature connection and a desire to uncover the seemingly impossible – a place where we can discover and live out our dreams, to completely indulge the mantra of 'Stop – Unplug – Escape – Enjoy'.

READER TESTIMONIALS

"When written well, traditional angling writing by the likes of BB, for example, is the type of literature that I can read again and again. Fennel's writing flows un-hurried without overly romanticising each point and the research is thorough; from the first sentence I was thinking, 'this lad can write!' It's informative and very refreshing."

"Such inspiring writing. His words 'Somewhere in the undergrowth of the impossible' had me staring out from the page in amazement. Fennel's writing is pure poetry."

From Fly Fishing:

"The deeper we travel into the natural world, and the greater the number of technological encumbrances we leave behind, the more likely we are to escape the fast-paced lifestyle and stresses of the 21st Century. For some, angling enables a quest into the unknown, an adventure into the wild. For these fortunate folk, fly-fishing is escapism. Their hours by water serve as contemplation to enrich their souls, directing their quest inwards, towards their longed-for state of completeness."

NO. 5

FLY FISHING

Fly Fishing celebrates the most graceful and artful form of angling, explaining what it means to be an angler – in the spirit of Izaak Walton – and how fly fishers differ from bait fishers. The sporting and aesthetic beauty of fly-fishing is described in Fennel's usual witty and contemplative style. As he says, "Fly fishing is the ultimate form of angling; it gives us a reason to fish simply, travel lightly, and explore wild places that replenish our soul. With a fly rod, we're not casting to a fish; rather to a circle of dreams: ripples that spread into every aspect of our lives".

READER TESTIMONIALS

"Brilliant writing. Fennel made me laugh out loud in bed. My wife was asking questions!"

"A delightful, well-articulated, read. I strongly recommend it, especially to the contemplative, tradition-loving, bamboo fly rod devotees among us."

"A very inspiring and rewarding read. I will try to tie the Sedgetastic fly. It looks tasty!"

From Traditional Angling:

"Physics teaches us that for every action, there is an equal and opposite reaction: a natural balance of energy that sustains the equilibrium of life. In modern angling, these forces are skewed so far in favour of technology that the balance between science and art has been lost. But there is a movement, an undercurrent that defies the flow of progress. There are those who choose not to follow the crowd. They seek not to fish in a predictable, scientific manner. They yearn for the opposite, to buck the trend, *to be different*. They are the Traditional Anglers."

NO. 6

TRADITIONAL ANGLING

Traditional Angling celebrates the Waltonian values of angling: about fishing in a seasonal and uncompetitive way for the pure pleasure of being beside water. It wears its heart on its sleeve and a wildflower in its lapel. It's passionate, provocative and eccentric, written for those who appreciate the aesthetics of angling and uphold its sporting traditions. So, with great enthusiasm, raise your bamboo rod aloft for an adventure that proves there's more to fishing than catching fish.

READER TESTIMONIALS

"A beautifully written, very engaging and hugely enjoyable read. In fact, it's the best thing on fishing I've read in a long time."

"What a Journal! Fennel is clearly the spiritual successor to his mentor – the great Bernard Venables. There's so much wisdom and craftsmanship in his writing. Bernard clearly taught him very well."

From The Quiet Fields:

"The countryside, with its vast horizons, fresh air and ever-changing seasons is, by its very nature, more life-giving and adventurous than any amount of modern indoor living. It inspires a love of natural history – everything from the birds that sing in the trees to the quality and richness of the soil beneath our feet. Most of all, it creates the desire to exist more naturally. And in doing so, we appreciate the balance of life."

NO. 7

THE QUIET FIELDS

The Quiet Fields is rooted in the humus-rich soil of the countryside. It's about remote rural places where Nature exists undisturbed, where we may sit and ponder 'The Wonder of the World'. The Journal tips its hat to these places, and to the nature writing of BB, revealing the 'Lost England' that still exists if you know where and how to look. It is the most sentimental and astutely observed Journal to date, discussing the 'true beauty' of Nature. If you've ever yearned to hear birdsong during a busy day, then this is the book for you.

READER TESTIMONIALS

"Fennel's writing reminds me of the works of Roger Deakin. It inspires me with faith in the quiet life and that although I may be isolated, I am certainly not alone."

"Fennel has captured the essence of the countryside – that is, its almost human character. So brilliantly has he compared and contrasted it with the nature of we humans. It's not so much a 'balanced study', more a 'study of the balance' between Nature and Man."

From Fine Things:

"It seems that, depending upon which side of the thesaurus-writer's gaze we sit, one's uniqueness as a person can be deemed to be either eccentric or distinctive. Both, in my opinion, are good...As we get older, and experience more things, those of us with strength of character and a sense of purpose will grow stronger and fight harder; those who lack identity and direction might end up sitting in a corner somewhere, blindly taking all the knocks that life throws at them. What does this teach us? That character and purpose are directly linked to confidence and conviction. What links them? Courage – to be oneself, no matter what others might say."

NO. 8

FINE THINGS

Fine Things celebrates the special and sentimental items and activities that convey our personality. The writing is fast-paced, quirky and humorous, reflecting the author's enthusiasm and eccentric view of the world. But be warned: if you look inside Fennel's mind, you might see a hula-hooping hamster named Gerald, shaking his maracas, loudly banging a bongo, and getting him into all sorts of trouble. So strap yourself in. This book picks up pace and takes some unexpected turns. From the deeply personal to the outright eccentric, it's for those who seek to be different.

READER TESTIMONIALS

"A very fine thing, indeed. Fennel's best and funniest book to date. He is the only author who can make me laugh out loud and cry in the same sentence. I was constantly in tears, for all the right reasons."

"Deep in places, outright bonkers in others. A demonstration of the fine line between genius and madness."

From A Gardener's Year:

"Roll up your sleeves and imagine your vision of paradise. This, in whatever form it takes, is your garden. Keep hold of the image; know it's every detail and piece together the elements that need creating or nurturing, so that when you get the chance, you can prepare the ground, sow the seeds, and make it real. Ours is a gardener's life, whether we realise it or not."

NO. 9

A GARDENER'S YEAR

A Gardener's Year celebrates the joy of growing things and reflects upon a life working with plants. But it's not a record of horticultural activities through the seasons. It's a metaphor for having a dream and making it come true. For Fennel, who has spent half his life working in gardens, it's about cultivating a cottage garden where he can aspire to a self-sufficient lifestyle. The Journal sees him sow the seeds of this future reality.

READER TESTIMONIALS

"Fennel's writing is uniquely funny. I mean, who else can name a chapter 'Chicken Poo'? His sense of humour, balanced with some deep yet subtle messages, had me in tears. From his 'escape' to a public toilet, to what not to say to a celebrity, this is a Journal to entertain all readers."

"When I started reading this Journal I had a garden with a lawn and a patio. Now I have a vegetable patch, blisters, an aching back, and the biggest smile of my life. Thank you Fennel!"

From The Lighter Side:

"If self-actualisation is the pinnacle of one's development, then it can't be achieved if your mountain has two peaks...Being the 'best version' of yourself implies that you have other versions kept locked in a closet. Don't have any 'versions'. Just have one true, beautiful and pure form of you.
So climb your mountain, open your arms to the Creator who greets you there, and sing loudly to the world that stretches out beneath you.
Write your name permanently on the landscape of your mind.
Remember: you are a child of Nature.
And you are free."

NO. 10

THE LIGHTER SIDE

There's a delicate balance between something meaning a great deal and that same thing becoming so serious that it's ludicrous. (Ever got stressed about what clothes to wear for an interview?) That's why *The Lighter Side* provides the encouragement, humour, anecdotes, reflections and honesty that are essential to Fennel's message of 'Stop – Unplug – Escape – Enjoy'. After all, we can only 'Enjoy' if we know how to smile when we get there.

READER TESTIMONIALS

"The Lighter Side was more than I expected. The deeper meaning within it – and the devastating honesty it conveys – made me question exactly where I am in my own life and what I can do to improve it for my family and me in the time that remains. Thank you Fennel for opening my eyes and adjusting my course."

"The opening chapter is the most startling, erudite, compassionate and open piece of writing I have ever read...thank you Fennel for sharing so much. It did and does mean a great deal."

From Friendship:

"What I'm talking about is proper friendship. The sort that is authentic, genuine and real. Where we can look into the eyes of another person and know what they're thinking. ...Because, as friends, we remember 'why' as much as 'when' or 'what'. Through good times and bad, we were there. Together. That's the bond, the unquestionable obligation that's freely given. It's the tightest hug, the biggest kiss, the tearful hello and the widest smile. If that's what it means to be a friend, or an extrovert, or just someone who cares for others then that's me to the last beat of my heart."

NO. 11

FRIENDSHIP

Written by the Friends of the Priory, with bonus chapters from Fennel, *Friendship* provides insights into what it means to be friends, how shared interests and beliefs support collective purpose, and how, when we're together, we can achieve more, appreciate more, and have more fun. It's about the broader world of Fennel's Priory and how it exists in others. It's a book written 'for us by us', with friendship as the theme.

READER TESTIMONIALS

"Possibly the greatest gift that this Journal bestows is to let us know that we are not alone."

"Like friendship itself, this Journal brings together people and meaning. It reminds us that 'together we are strong'. Thank you Fennel for leading our charge."

"The message (and evolution) of Fennel's Journal is most evident in this Friendship edition. With such obvious themes as identity and legacy, it's clear that what Fennel has shared over the years is a route-map to freedom and a stronger sense of self."

From Nature Escape:

"I am once again seeking an escape, to where I hope to find freedom and connect with the young man who handed me his trust ten years ago. This will be a faithful interpretation of the Priory, a fitting way to mark ten years of writing. As I said at the end of last year's Journal, 'One's journey through life is not linear; it's circular.' So let's go back to the beginning, and rediscover the quiet world."

NO. 12

NATURE ESCAPE

Nature Escape provides the most detailed account of a day that follows the motto of 'Stop – Unplug – Escape – Enjoy'. In it Fennel returns to the woodland of his youth to study its wildlife and savour its peacefulness.

Written in real-time, with twenty-four chapters that each represent an hour, the Journal is an account of how time spent outdoors in wild places enables us to observe the nature that's around us *and* within us.

READER TESTIMONIALS

"Fennel's Journal has always provided us with an escape, but now we know where the escape can lead. As promised, it leads to enjoyment – and very enjoyable it is too!"

"24 hours alone in a wood, with only 'the wild' for company? With Fennel as our guide, there's no such thing as 'alone'; only the warmth of knowing that quiet times are the fine times."

"By studying the nature within us and around us, Fennel demonstrates how to be 'at one' with nature."

From Book of Secrets:

"There's a greater man than me who can sum up our journey, a mountaineer who in 1865 first climbed the Matterhorn. Edward Whymper, over to you: 'There have been joys too great to be described in words, and there have been griefs upon which I have not dared to dwell, and with these in mind I say, climb if you will, but remember that courage and strength are naught without prudence, and that a momentary negligence may destroy the happiness of a lifetime. Do nothing in haste, look well to each step, and from the beginning think what may be the end.'"

NO. 13

BOOK OF SECRETS

Book of Secrets links all editions of Fennel's Journal together. With 14 Journals in the series, and 14 core chapters in this book, it's the 'one book to bind them all' with each chapter providing the continuity story from one Journal to the next.

Containing Fennel's previously private writing, it provides deep insight into the Fennel's Journal story. If you've ever wondered why each Journal is themed the way it is, or tried to find the metaphor in each edition, then *Book of Secrets* is for you.

READER TESTIMONIALS

"What a privilege: being able to read the private writing of my favourite author. Book of Secrets is a treat."

"Such honesty and wit. Fennel puts into words what I have only ever thought, or dare not say."

"Fennel's Journal really is a series – it's meant to be read as a whole. And now we have the key to unlock it."

From The Pursuit of Life:

"We can hide, or we can strive – for a life of our making. With endless possibilities and opportunities to reach for our dreams, we owe it to ourselves to dream big and keep going, irrespective of what we might encounter. Sadly, the thing that most limits our success is not others, but ourselves. How strongly we believe, how confidently we act, how fiercely we react, how passionately we want, and how life-affirmingly compelled we are to grow and blossom; that's how we keep going, no matter what, to be the person we want to be, living the life we deserve, in dreams that are real."

NO. 14

THE PURSUIT OF LIFE

The Pursuit of Life concludes the Fennel's Journal story. It's a reflective tome that provides Fennel's commentary on the journey and a 'behind the scenes' view of the challenges and rewards of a life rebuilt on one's terms.

It's an account of how the series came to be and how it evolved, and includes much of Fennel's private writing, several of the original handwritten drafts, correspondence between The Friends, and encouragement for those on similar paths. Ultimately it shows how the Fennel's Journal series can be used as a route map to a more fulfilling life.

READER TESTIMONIALS

"A life retold, for our benefit. Fennel is to be congratulated for everything he's achieved – on paper and in life."

"It's his life in the books, but it could so very easily be ours. Fennel has a way of seeing truth in the severe and the sublime, and bringing it home."

"Can this really be the end? When dreams are real, we never wake from them. More books Fennel, please!"

www.ingramcontent.com/pod-product-compliance
Lightning Source LLC
Chambersburg PA
CBHW030521310726
48979CB00010B/1753/J

* 9 7 8 1 9 0 9 9 4 7 5 8 0 *